Coconut Palm Sap

A Natural Beverage and Source of Value-Added Products

nova
science publishers

www.novapublishers.com

Horticulture, Viticulture and Viniculture Series

Pomegranate: For Horticulture Students and Farmers
Arkalgud N. Ganeshamurthy, PhD
eBook ISBN: 979-8-88697-812-4

Ficus carica: Production, Cultivation and Uses
Zeynel Dalkılıç, PhD (Editor)
ISBN: 978-1-68507-765-5
eBook ISBN: 978-1-68507-873-7

Vitis: Biology and Species
António Manuel Santos Tomás Jordão and Renato Vasconcelos Botelho (Editors)
ISBN: 978-1-53618-308-5
eBook ISBN: 978-1-53618-424-2

More information about this series can be found at
https://novapublishers.com/product-category/series/horticulture-viticulture-and-viniculture/

K. B. Hebbar
V. Krishnakumar
S. V. Ramesh
Editors

Coconut Palm Sap

A Natural Beverage and Source of Value-Added Products

nova
science publishers
www.novapublishers.com

DOI: https://doi.org/10.52305/EJMS4171

Library of Congress Cataloging-in-Publication Data

ISBN: 979-8-89530-640-6 (Hardcover)
ISBN: 979-8-89530-705-2 (eBook)

Published by Nova Science Publishers, Inc. † New York

Contents

Preface

Palms are among the best-known and most extensively cultivated plant species, native to tropical and subtropical regions around the world. The evolution of research into palm crops is deeply rooted in the culture and history of the regions where these crops are grown. Furthermore, the research and development of these plantation crops cannot be viewed as an isolated component, given the intricate relationships these palms weave into the social fabric of these regions. One of the economic applications of palms involves the proper tapping technique to extract sugary phloem sap, which is characterized by high nutritive value. The inflorescences of palms like coconut, palmyra, and fish tail palms are tapped to obtain sap, while in date palms, the stem is tapped. In oil palms, both the inflorescence and stem are tapped. The rationale behind tapping palms for sap or sugar production is to intercept the sugar in the phloem sap and process it externally before it is diverted to the production of uneconomic parts such as husk and fibers, thus improving energy use efficiency.

In coconut, sap is tapped from unopened spadices. The phloem sap is rich in sugars, proteins, amino acids, minerals, vitamins, and phenolics, offering immense antioxidant potential and making it a healthy alternative to other plant juices. Unfermented sap is consumed directly as a health beverage, and it can also be concentrated to produce syrup, jaggery, sugar, and other value-added products. However, traditional methods of sap collection lead to rapid fermentation, resulting in the production of toddy, which renders the sap unfit for consumption as a non-alcoholic beverage or for use in sugar production. The advent of scientific sap collection methods using coco sap chillers is considered a game-changer in the palm industry, as it allows for the extraction of unfermented, hygienic palm sap with no alcohol traces, unlocking the potential to create various value-added products.

This book introduces the concept of palms and sap-yielding palms, detailing their prevalence and utilization. It also explores the effect of coconut genotypes and seasons on sap yield, followed by a chapter discussing environmental factors, irrigation, and nutritional requirements that influence palm sap yield. There is a dedicated chapter on the biochemical composition and quality attributes of coconut palm sap, as well as one on advancements in sap collection methods. Further chapters cover fermentation chemistry, preservation methods, and the storage and shelf life of coconut palm sap. This sets the stage for chapters on coconut palm sap as a beverage, raw material for producing other value-added products, and the nutritional profiling of coconut sugar and related products. Another chapter explains the health and nutritional potential of coconut palm sap and its value-added products. To provide insight into economic aspects, a chapter discusses consumer preferences, as well as the potential for income and employment generation. Finally, the concluding chapter describes the future prospects of coconut palm sap.

It has been a great privilege to collaborate with experts in palm sap characterization and utilization, food processing, biochemistry, physiology, and horticulture, among others, in preparing this book. We are deeply grateful to the authors and peer reviewers for their contributions, time, and expertise. As editors, we hope this book serves as a pioneering and comprehensive resource on palm sap and its products, offering valuable insight to palm scientists, scholars, and entrepreneurs. It also highlights the nutritional and health benefits of palm sap and products, while exploring opportunities for farm income generation. In summary, this book will undoubtedly serve as an essential reference for researchers working in the field of palms and palm-based food products.

K. B. Hebbar, Kasaragod, Kerala
V. Krishnakumar, Kayamkulam, Kerala
S. V. Ramesh, Kasaragod, Kerala

Acronyms

ABV	Alcohol by volume
ADH	Alcohol dehydrogenase
AFS	Anti-fermenting solution
AI	Artificial Intelligence
APCC	Asian and Pacific Coconut Community
AGD	Aromatic Green Dwarf
$_{a}W$	Water activity
BAFS	Bureau of Agriculture and Fisheries Standards
CAGR	Compounded annual growth rate
CAMPCO	Central Arecanut and Cocoa Marketing and Processing Cooperative L
CATD	Catigan Dwarf
CDB	Coconut Development Board
CDs	Carbon dots
CE	Catechin Equivalent
CFTRI	Central Food Technological Research Institute
COD	Chowghat Orange Dwarf
CPCs	Coconut Producer Companies
CPFs	Coconut Producer Federations
CS	Cane Sugar
CSCM	Coco-Sap Chiller Method
CSP	Coconut Sap Powder
CVD	Cardiovascular Disease
DFRL	Defence Food Research Laboratory
DPPH	2,2-diphenyl-1-picrylhydrazyl
DRDO	Defense Research and Development Organization
ED-XRF	Energy-dispersive X-ray fluorescence
EFSA	European Food Safety Authority
EGD	Equatorial Green Dwarf
FAA	Free Amino Acids
FCIS	Fresh coconut inflorescence sap

FDA	Food and Drug Administration
FNRI	Food and Nutrition Research Institute of India
FSSAI	Food Safety and Standards Authority of India
GAE	Gallic acid equivalent
GCS	Granulated Coconut Sugar
GI	Glycemic Index
GMP	Good manufacturing practices
GOA	Grand Ouest Africain
HDL	High-density lipoprotein
HMF	5-hydroxyethylfurfural
hPBMCs	human peripheral blood mononuclear cells
ICC	International Coconut Community
II	Insulinemic Index
IoT	Internet of Things
IRMS	Isotope Ratio Mass Spectrometry
LAB	Lactic acid bacteria
LAGT	Laguna Tall
LC-MS	Liquid Chromatography-Mass Spectrometry
LCT	Laccadive Ordinary Tall
LDL	Low-density lipoprotein
MD	Maltodextrin
ML	Machine Learning
MR	Maillard reactions
MRD	Malayan Red Dwarf
MSMEs	Medium and Small Micro Enterprises
MYD	Malayan Yellow Dwarf
NCIS	Naturally fermented coconut inflorescence sap
NMR	Nuclear Magnetic Resonance
NO	Nitric oxide
NPNC	Non-alcoholic palm nectar
NPV	Net present value
NYD	Nias Yellow Dwarf
OHE	Open Heat Evaporation
OxLDL	Oxidized low-density lipoprotein
PET	Polyethylene Terephthalate
PL	Pulsed light
PNS	Philippines National Standard
PVPP	Polyvinyl Poly Pyrrolidone
RBD	Raja Brown Dwarf

RSA	Radical Scavenging Activity
RTS	Ready to Serve
SCTNP	Sequential Coconut Toddy and Nut Production
SGD	Salak Green Dwarf
SGD	San Isidro Green Dwarf
SNI	Indonesian National Standard
STZ	Streptozotocin
TACD	Tacunan Dwarf cultivar
TAL	Talisay
TBARS	Thiobarbituric acid-reactive substance
TCC	Total carbohydrate contents
TCP	Tricalcium phosphate
TE	Trolox equivalent
Tg	Glass transition temperature
TSS	Total soluble Solids
UHT	Ultra-High Temperature
VCO	Virgin Coconut Oil
VOCs	Volatile organic compounds
WAT	West African Tall
WCT	West Coast Tall
WHO	World Health Organisation
WRD	Waingapu Red Dwarf

Chapter 1

Palms and Sap Yielding Palms - Their Prevalence and Utilization - Global Scenario

B. A. Jerard[1,*]
S. Sumitha[1]
S. V. Ramesh[2]
V. Krishnakumar[3]
and K. B. Hebbar[2]
[1]AICRP on Palms, ICAR-Central Plantation Crops Research Institute, Kasaragod, Kerala, India
[2]ICAR-Central Plantation Crops Research Institute, Kasaragod, Kerala, India
[3]ICAR-Central Plantation Crops Research Institute, Kayamkulam, Kerala, India

Abstract

Palms (Arecaceae family) are tropical and subtropical plants known for their ecological and economic significance. They are prevalent in regions like Southeast Asia, Africa, and Latin America, where they play vital economic roles. Palms such as the coconut palm (*Cocos nucifera*), date palm (*Phoenix dactylifera*), and sugar palm (*Arenga pinnata*) are harvested for their sap, which can be used to produce sweeteners (like palm sugar) and alcoholic beverages (like palm wine). Some palms, like the nipa palm (*Nypa fruticans*), also provide biofuel through ethanol production. Globally, palms play important role in food security, energy production, and cultural practices. In South and Southeast Asia, palm sap is used in the production of sugar and vinegar, while in Africa, palm wine is a popular local beverage. These palms are also considered as a major

[*] Corresponding Author's Email: jerardba@gmail.com

In: Coconut Palm Sap
Editors: K. B. Hebbar, V. Krishnakumar and S. V. Ramesh
ISBN: 979-8-89530-640-6

component of sustainable agriculture and renewable resources. This chapter discusses geographic spread of palms, tapping of palms, nutritive value and economic importance of palm sap and also provides glimpses of value-added products from palm saps.

Keywords: sap-yielding palms, traditional uses of palms, palm sap harvesting, sustainable palm agriculture, palm sugar

1. Introduction

Palms are one of the world's oldest flowering plants known to mankind. Generally, most useful palms are known as 'Tree of Life' and are extensively found and cultivated in the tropical and sub-tropical regions of the world (25°Nand 25°S of the equator). Palms are monocotyledonous angiosperm that belongs to Arecaceae family with 2500-2700 known species including six sub-families and 200 genera.

Palms such as *Cocos nucifera* (coconut palm), *Arenga pinnata* (sugar palm), *Hyphaene* spp. (real fan palm, Ivory palm), *Nypa fruticans* (Nipa palm), *Elaeis guineensis* (African oil palm), *Phoenix sylvestris* (sugar date palm), *Corypha umbraculifera* (Talipot palm), *Raphia* spp. (Raffia palm, Wine palm, Bamboo palm), *Caryota urens* (kitul palm), *Borassus flabellifer* (Palmyra palm), *Phoenix dactylifera* (date palm), *Areca catechu* (betel nut palm) and *Bactris gasipaes* (Pejibaye - peach palm) are of greater economic importance owing to their diversified applications in several industries such as coir, oil, sago and sugar (Hebbar et al., 2018; Jayanudin et al., 2019; Shetty et al., 2017; Van Nguyen et al., 2016).

For many centuries, various palm species have been tapped throughout the tropical world to produce fresh juice (sweet toddy), fermented drinks (toddy, wine, arak), syrup (honey), brown sugar (jaggery) or refined sugar (Figure 1). One of mankind's earliest sources of sugar was likely *Arenga pinnata*. Evidence of the use of *Borassus flabellifer* sugar in India has been reported by the Greek historian Megasthenes, ambassador to the court of Chandragupta in the 4th century BC. Indians knew how to extract palm sugar about 4,000 years ago. Jaggery and treacle produced from sap of *Caryota urens* in Sri Lanka has been an important source of sugar since antiquity (Dissanayake 1977). In Africa, the primary traditional use of palm sap is wine production. The food use of date palm has been reported in Egypt since time immemorial (Barreveld 1993) and along the Guinea coast by early navigators in the 15th century (Ayernor and Matthews 1971). The collected sap is honey to oyster white in colour, translucent, sweet in taste and is widely consumed in the rural regions (Sudha et al., 2019). The sap is commonly known as '*neera*' '*pathaneer,*' '*Kalparasa*' in India and is released from the phloem in a sieve tube system (Borse et al., 2006; Van Nguyen et al., 2016). It is also known as tuba, nectar or honey water in Colima, Mexico and Guarapo in Canary Island (Chandrashekhar et al., 2012; Flores-Galleges et al., 2019). It is a nutritious fluid enriched with both macro and micronutrients and sugars, minerals, antioxidants, and vitamins. It is classified as foods that have low glycemic index (GI 35) (Ghosh et al., 2018; Hebbar et al., 2018; Trinidad et al., 2010). However, another group of researchers have reported a GI level of 52.17 for coconut sap which is in intermediate range (Chayanika et al., 2021). GI of coconut sap may vary due to the age of the tree, variety, climate and geographical location, and hence a detailed study is warranted to establish the accurate GI of palm saps (Chinnamma et al., 2019).

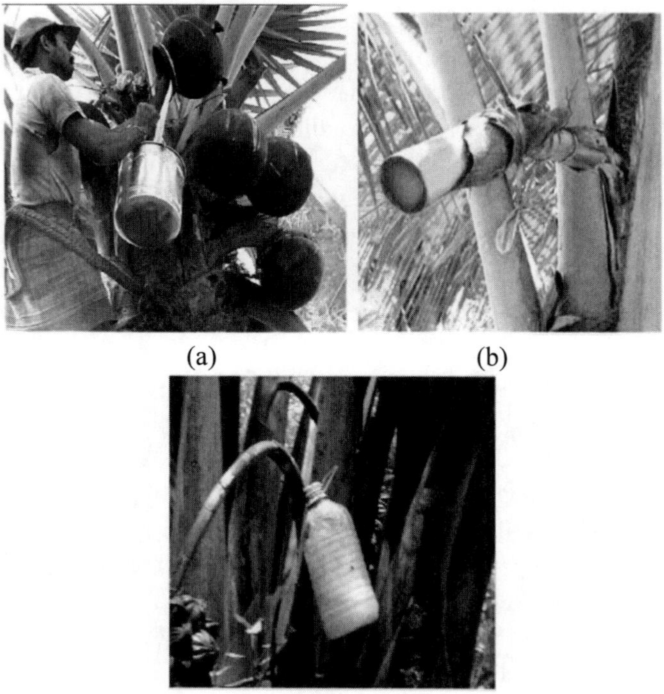

(a) (b) (c)

Figure 1. Tapping of different palms a) Palmyrah palm b) Coconut palm c) Nipa palm.

The most appealing feature of the sap is its flavour and nutritional components. This organic sap is reported to have various health benefits *viz;* anti-diabetic, anti-ageing, anti-rheumatic, anti-electrolyte imbalance (Hebbar et al., 2015). Thus, this thirst-quenching sap is gaining demand due to its natural origin and high nutritive value (Kapilan 2015). Though the sap has high nutritive value and health benefits, there is lack of awareness among people about its potential nutritional benefits. The sap is consumed as a traditional drink in rural areas. Due to the lack of proper preservation techniques that limit its transportation process, the urban population is unaware of its health prospects. The presence of diverse microflora in the sap naturally warrants it preservation. Further, the lack of technology for hygienic tapping is another major drawback that severely limits the shelf life of the sap. The freshly tapped sap must be processed immediately as it gets fermented quickly, which alters the taste and quality. The fermented sap has a pungent odour and acidic taste due to the increase in the acid content, ethyl esters and ethyl alcohol (Borse et al., 2006). Traditional tappers generally use lime coating on the inner lining of the collection pot to inhibit fermentation. Several other methods have been attempted to inhibit the fermentation process, but those methods are less successful (Hebbar et al., 2018). It is suggested to implement hygienic measures in handling during harvest, processing and storage as the sap is extremely prone to instant fermentation. Since rapid fermentation of the sap makes the preservation a challenging task, it is converted to value-added products and marketed globally on a commercial scale.

2. Geographic Spread, Products and Tapping of Palms

Table 1 enlists nearly 30 different palm species that are tapped and found adapted to almost all agro-ecological zones of the tropical world from tidal areas and swamps to deserts and mountains. Majority of the information pertains to palms tapped in the Old World, with as many different tapped species in Asia and in Africa. Only three tapped palm species have been identified in the New World: *Carnauba cerifera*, *Jubaea spectabilis*, and *Mauritia flexuosa*. In the Americas and Africa, palm tapping hase been practiced exclusively or primarily for wine production. Contrarily, in Asia, the sap is utilized either as fresh juice or processed into a wide variety of products, including wine, arak, sugar, vinegar, etc.

Table 1. Worldwide distribution of palms, tapping parts, tapping method, yield and products

Scientific name (common name)	Distribution	Palm part and method of tapping	Sap yield (l $palm^{-1}day^{-1}$)	Products	Age of first sap harvest (years)	Tapping Season	Reference
Acrocomia aculeate (Macaw palm, Coyol palm)	Tropical regions of the Americas (Mexico, Caribbean countries, Paraguay, Argentina)	Terminal bud; Destructive	2	Wine	-	August to December	Francisco-Ortega and Zona (2013)
Acrocomia mexicana	Honduras	Terminal bud on felled trees; Destructive	2	Wine	5	Year round	Lentz (1990)
Areca catechu (Areca palm, betel nut palm, arecanut)	Tropical rain forest to South and Southeast Asia: India, Indonesia and East Africa	Male (and rarely female) inflorescence spadix or secondary stems; Non-destructive	1.5-3	Sugar (jaggery), fresh sap, toddy	7-8	Year round	Staples and Bevacqua (2006)
Arenga pinnata (Sugar palm)	Humid areas of tropical South and Southeast Asia (India, Sri Lanka, Guam, Papua, New Guinea, Indonesia, Thailand, Vietnam)	Stalk; Non-destructive	3-6	Sugar (jaggery), fresh sap, toddy, vinegar	5–12	Year round	Victor (2015)
Arenga wightii (Wight's sago palm)	India	Inflorescence (spadix); Non-destructive	2	Toddy	5–6	April	Pillai et al., (2020)
Attalea butyracea (Yagua palm)	The Caribbean lowlands, the Eastern Plains and the Amazon region, Mexico throughout Central America and into Colombia, Venezuela, Peru, Brazil, and Bolivia	Crown meristem of felled palm; Destructive	1–3.7	Fresh sap, wine, Vinegar	15-25	Year round	Bernal et al., (2010)

Table 1. (Continued)

Scientific name (common name)	Distribution	Palm part and method of tapping	Sap yield (l $palm^{-1} day^{-1}$)	Products	Age of first sap harvest (years)	Tapping Season	Reference
Beccariophoenix madagascariensis	Central Madagascar	Trunk; Non-destructive	-	Fresh sap, Juice, Wine	-	-	Van Nguyen et al., (2016)
Borassus aethiopum (African fan palm)	Tropical zone from West Africa through India and Southeast Asia to New Guinea and Australia	Palm heart (apical meristem); Destructive	10	Wine, fresh sap, vinegar, potential for sugar production	20	April to October, Rainy season	Dalibard (1999); Sambou et al., (2002)
Borassus akeassii	Sub-Saharan Africa (Senegal, Mali, Ivory Coast, Niger and Burkina Faso)	Stem below terminal bud; Non-destructive	0.5-10	Fresh Sap	22	November to May	Tapsoba et al., (2016)
Borassus flabellifer (Palmyra palm/ Lontar palm)	Tropical countries in Asia (Nepal, Sri Lanka, India, Malaysia, Indonesia, the Philippines, Vietnam)	Inflorescence (spadix); Non-destructive	6–10	Sugar candy, jaggery, toddy, neera, vinegar	20-30	February to March (Dry season)	Van Nguyen et al., (2016)
Borassus madagascariensis	Madagascar	Inflorescence spadix; Destructive	1	Wine	-	-	Decary (1964)
Calamus vanuatuensis	New Guinea, The Solomon Islands, Va	Stem; Non- destructive	1.5–3	Sap	-	November to January	Dalibard (1999); Dowe (2010)
Caryota mitis (Clustering fishtail palm)	India, Brunei, Malaysia, Myanmar; Indonesia, Thailand, Vietnam	Inflorescence; Non-destructive	1.5–3	Sap	15–30	-	Van Nguyen et al., (2016)
Caryota urens (Indian sago palm, Kitul Palm)	Humid areas of South Asia (India, Sri Lanka, Malaysia, Indonesia, Philippines)	Stalk (peduncle); Non-destructive	45	Sugar, jaggery, fresh sap, toddy	10–15	October to June	Van Nguyen et al., (2016); Patel and Neema (2021)

Scientific name (common name)	Distribution	Palm part and method of tapping	Sap yield (l $palm^{-1} day^{-1}$)	Products	Age of first sap harvest (years)	Tapping Season	Reference
Cocos nucifera (Coconut Palm)	Common to tropical lands	Inflorescence (spadix); Non-destructive	1.7–4.3	Sugar, sap, toddy	5	Year round	Hebbar et al., (2015) Van Nguyen et al., (2016)
Corypha elata (Bajur palm, Buri palm, sabal palm, Sulphi palm)	Southeast Asia: Indonesia, Philippines	Immature inflorescence; Non-destructive	20	Fermented drink, vinegar, muscovado sugar	30–60		Dalibard (1999)
Corypha umbraculifera (Talipot palm)	Tropical rainforest of South and Southeast Asia (Sri Lanka, India, Myanmar, Thailand, Cambodia)	Inflorescence; Non-destructive	20	Sugar, wine	20	-	Irawanto (2013)
Corypha utan (Buri Palm)	Wide distribution in dry and open areas of Asia (India, Sri Lanka, Bangladesh, Malaysia, Indonesia, Philippines, Australia)	Inflorescence; Non-destructive	45	Sap	30–60	-	Irawanto (2013)
Elaeis guineensis (Oil palm)	Tropical rain forest regions of Africa, Southeast Asia, South and Central America (Nigeria, Ivory Coast, Cameroon, Madagascar, Angola, Malaysia, Indonesia, Colombia)	Terminal bud – Destructive; Felled trunk-Destructive; Inflorescence (spadix) - Non-destructive	4-5	Wine, strong alcohol	6–10	-	Joseph et al., (2018)
Hyphaene coriacea (Lala palm)	Arid parts of Africa (Madagascar, South Africa)	Terminal bud; Destructive	4.4–14.5	Wine		Year round	Francisco-Ortega and Zona (2013); Martins and Shackleton (2018)

Table 1. (Continued)

Scientific name (common name)	Distribution	Palm part and method of tapping	Sap yield (l $palm^{-1}day^{-1}$)	Products	Age of first sap harvest (years)	Tapping Season	Reference
Hyphaene petersiana (Real fan palm, Ivory palm)	Subtropical, low-lying regions of South-Central Africa and Zimbabwe	Terminal bud; Destructive	1	Wine	-	February to April	Babitseng and Teketay (2013); Sola et al., (2006)
Hyphaene thebaica (Doum Palm)	West from Mauritania and Senegal, and east to Egypt, Kenya and Tanzania, Nile basin in Sudan	Apical meristem; Destructive	4	Wine	6–8	November to April	El-Beltagi et al., (2018); Van Nguyen et al., (2016)
Jubaea chilensis (Chilean palm)	South America (Chile)	Apical meristem of uprooted palm - Destructive Apical meristem - Non – destructive	8	Wine, Palm Honey	5–15	Summer December to February	Gonzalez (2009)
Mauritia flexuosa (Buriti palm)	Near swamps and other wet areas in tropical South America (Trinidad, Colombia, Venezuela, Guyana, Suriname, French Guinea, Brazil, Ecuador, Peru, Bolivia)	Terminal bud - Destructive Inflorescence - Non- destructive	1.5–3	Sugar, Wine	12-20	Year round	Francisco-Ortega and Zona (2013); Delgado et al., (2007)
Metroxylon sagu (Sago palm)	Humid tropical lowlands, up to an altitude of 700 m (Papua New Guinea, Melanesia, Indonesia, Malaysia, Thailand)	Stalk; Destructive	2–10	Toddy	15	-	Corbishley and Miller (1984)

Scientific name (common name)	Distribution	Palm part and method of tapping	Sap yield (l $palm^{-1} day^{-1}$)	Products	Age of first sap harvest (years)	Tapping Season	Reference
Nypa fruticans (Nipa palm)	Soft mud and slow-moving tidal areas such as coastlines, estuaries, mangrove forests (India, Sri Lanka, Bangladesh, Burma, Thailand, Cambodia, Malaysia, Indonesia, Philippines, Vietnam, Nigeria	Stalk (cut off inflorescence); Non-destructive	1.3	Fresh sap, toddy, sugar and vinegar	5	Year round	Matsui et al., (2014)
Orbignya martiana (Babassu Palm)	Wide range of edaphic and climatic conditions, Tropical Latin America, Brazil	Stump hollowed out after felling the tree; Destructive	1.5–3	Wine	10–20	Year round	May et al., (1985)
Phloga polystachya	Madagascar	Trunk; Destructive	1.5–3	Fresh sap, wine	-	-	Decary (1964)
Phoenix canariensis (Canary Island date)	Canary Islands and Southern California	Apical meristem; Non-destructive	1.5–3	Fresh sap	5–10	Summer	Meekijjaroenroj (2004)
Phoenix dactylifera (Date palm)	Arid and semiarid regions of western Asia and North Africa (Egypt, Iran, Saudi Arabia, United Arab Emirates, Pakistan, Algeria, Iraq, Sudan, Oman, Libya)	Growing point of palm (terminal bud), Non-destructive	8–15	Wine	Above 5	March to June	Al-Alawi et al., (2017): Makhlouf-Gafsi, Ines et al., (2016)

Table 1. (Continued)

Scientific name (common name)	Distribution	Palm part and method of tapping	Sap yield (l $palm^{-1} day^{-1}$)	Products	Age of first sap harvest (years)	Tapping Season	Reference
Phoenix reclinata (Senegal date palm)	Tropical Africa, the Arabian Peninsula, Madagascar and the Comoro Islands	Terminal bud - Destructive Inflorescence - Non- destructive	1.5–3	Palm gur, wine		September to November (Zambia); August to October (South Africa)	Lim (2016); Orwa et al., (2009)
Phoenix sylvestris (Wild date palm)	Arid and desert areas of northern Africa, the Middle East and Southern Asia (Arabian Peninsula, Iran, Pakistan, Bangladesh, India)	Stem below terminal bud, Non-destructive	1.2–2.5	Wine, sugar	Above 4	August	Namoff et al., (2011)
Raphia farinifera	Wet areas of Madagascar	Terminal bud, Destructive	2	Fresh sap, wine, strong alcohol	Above 4	Year round	Dalibard (1999); Francisco-Ortega and Zona (2013)
Raphia hookeri (Raffia palm, Wine palm)	Swampy areas in forest regions of Africa (Nigeria, Madagascar, Ghana, Cameroon, Gabon, Congo)	Terminal bud – Destructive Inflorescence – Destructive	2	Fresh sap, wine, strong alcohol	Above 4	Year round	Francisco-Ortega and Zona (2013); Obahiagbon and Osagie (2007)
Raphia vinifera (Bamboo palm)	Tropical Africa (Cameroon)	Destructive		Fresh sap, wine, strong alcohol		Year round-	

3. Economic Importance of Palm Sap

The increased health awareness among people has created a demand for drinks of natural origin. In this context, palm saps are of great economic importance especially for rural and sub-urban population. The Philippines, Thailand, and Sri Lanka are the major manufacturers of palm sap from *Cocos nucifera* while Malaysia and Indonesia harvest mostly from *Nypa fruticans* (Van Nguyen et al., 2016), India and Sri Lanka from *Borassus flabellifer.* Sap is extensively exported to countries like Canada, South Korea, USA, Norway, France, Japan, Australia and the Middle East (Sudha et al., 2019). The global beverage industry is expanding rapidly, with revenue of $235.70 billion in 2024, at a compound annual growth rate (CAGR) of 10.05% between 2024 and 2029 (Reinoso 2024).

According to World Health Organisation (WHO 2011), 346 million people are diabetic and require an alternative sweetener, preferably of natural origin, to control the blood glucose level. As palm saps are categorized low GI foods, the market value of the sap as such or the value-added products may increase by several folds (Trinidad et al., 2010). Jnanadevan (2015) reported that a farmer's income can be raised by 15 folds from a coconut palm by tapping the sap. Annually, this nutritious sap holds a vital role in Indian economy by contributing $105 million to the country's GDP. A farmer owning 100 palms can generate employment in the local community with an average net return of INR 9.24 lakhs ($12,700) year^{-1} (Ghosh et al., 2018). In India, palm sap tapping and value-addition are major activities that support only the domestic market, while the global market is shared by countries like Indonesia (18%), Philippines (38%), Malaysia (8%), Sri Lanka (14%) and Thailand (22%) (Coconut Development Board, India).

The processing of palm sap could be one of the major drivers of industrialization in rural areas. Due to the lack of proper preservation technologies at present, huge quantity of palm sap is being wasted every year (Hebbar et al., 2018). Several attempts have been made to improve the shelf life of the palm sap by adopting different processing technologies such as freezing, freeze drying, spray drying, cold plasma, nanofibrous membrane filtration etc. However, multiple limitations such as low acceptability of the product or unavailability of commercial scale equipments to process persist (Hebbar et al., 2018; Leena et al., 2021; Sukumaran and Radhakrishnan 2021). Thus, appropriate marketing needs to be carried out according to the availability of sap from the geographical location. Depending upon the demography, sustainable technologies shall be adopted for improving the economic value which will be beneficial for food industries.

The management of palm trees for sap production varies depending upon the species. *Nypa fruticans*, *Phoenix sylvestris*, *Elaeis guineensis*, *Raphia hookeri* and *Cocos nucifera* can be tapped at an early age when the trees are 4, 5, 6, 7 and 7 years old, respectively. On the other hand, tapping *Caryota urens* (10 to 15), *Borassus flabellifer* (15 to 30) or *Corypha elata* (20 to 100) requires many years (Table 1). The number of years a palm tree can be tapped also depends on the species. *Corypha elata* and *Raphia hookeri* flower only once and are known to produce sap only for a few months before dying. *Arenga pinnata* and *Caryota urens* produce sap for several years, despite large interruptions. For instance, *Caryota urens* flowers only every two or three years. Other palm trees will produce sap for much longer periods: examples include 10 to 15 years for *Elaeis guineensis*, more than 20 years for *Cocos nucifera*, 50 years for *Nypa fruticans* and *Phoenix sylvestris* and 30 to 100 years for *Borassus flabellifer.* Some species produce sap all year round (*Arenga pinnata*, *Cocos nucifera*, *Elaeis guineensis* and *Nypa fruticans*) while others show seasonal production of sap *(Borassus flabellifer* and *Phoenix sylvestris*).

4. Tapping of Palms

Traditionally, palm sap is tapped either early in the morning or late in the evening by placing an earthen pot at the cut end of inflorescence. The yield varies depending on the genetic make-up of the selected tree (Misra 2016). Although various methods of tapping are practiced worldwide, the principle of sap collection remains similar. The sap yield may vary depending on the tapping method, skill of tapper, season and geographical location (Van Nguyen et al., 2016). Generally, tapping is carried out in three spathes on the same palm continuously for six months. A spray of ethrel is also practiced to increase the flow rate of the sap as it inhibits the formation of callus in the cut part of the spadix. On the other hand, disturbance in the phloem flow may result in the reduction of sap yield (Sudha et al., 2019). Generally, destructive or non-destructive techniques are adopted for tapping sap of palms.

4.1. Destructive Tapping

Destructive tapping is performed by slicing the stem or by puncturing the palm trees. In this method, the stem being tapped might die as the embryonic tissues are damaged while tapping. However, this method of tapping is suitable for clustering palms with clumping stems for example in a few species such as *Elaeis guineensis, Hyphaene petersiana, Hyphaene coriacea, Jubaea chilensis, Phoenix reclinata, Pseudophoenix ekmanii, Attalea butyracea, Borassus aethiopum, Pseudophoenix vinifera, Raphia farinfera, Raphia hookeri*. Apparently, this method is regarded as unsustainable as it is reported to be the causative factor for the extinction of *J. chilensis* and for the decline of *A. aculeate, B. aethiopum* and *B. madagascariensis* in Central America, Republic of Guinea and Madagascar, respectively (Table 1).

4.2. Non-Destructive Tapping

Non-destructive method is the most common form of tapping practiced in Asia and Africa which includes tapping of stalk, stem and unopened inflorescence following a pre-treatment like beating the stalk to trigger the sap flow. The head of the mature unopened inflorescence is cut for about 7–10 cm. It is practiced when swelling is observed at the base of the inflorescence (Hebbar et al., 2015; Hebbar et al., 2018). The spathe is then tied with a coir rope or palm fibre to avoid the inflorescence from burst opening. A palm can yield about 8–10 l of sap day^{-1} by following non-destructive tapping. Regrowth of apical meristem allows the palms to be tapped again for 3 to 4 times in every 5 years time period making it a sustainable approach (Jnanadevan 2015). This method is beneficial and practiced for tapping palms in different regions worldwide in species like *Arenga pinnata, Arenga wightii, Borassus akeassii, Borassus flabellifer, Nypa fruticans, Caryota mitis, Cocos nucifera, Corypha umbraculifera, Metroxylon sagu, Phoenix dactylifera, Phoenix canariensis* etc. (Van Nguyen et al., 2016) (Table 1).

5. Nutritive Value of Palm Sap

Nutritional composition of palm sap varies to a larger extent and depend upon the geographical region, genus, species, variety and duration of tapping (Borse et al., 2006). The sap is mainly composed of 80% water and 10–15% of total sugars with sucrose as the major constituent along with glucose, fructose, inositol and raffinose at varying levels. Sugar produced from *Borassus flabellifer* and *Phoenix sylvestris* sap was found to be healthier compared to cane sugar, as the protein and calcium content of these palm sugars are 3.63 and 1.32 times higher (Dalibard 1999; Jagannadha Rao 2009). Sucrose is the major constituent present in the sap of all palms. The highest sucrose content of 37.8% is found in the sap collected from *Phoenix canariensis* (Luis et al., 2012) while in other palm saps (*Borassus flabellifer, Nypa fruticans* and *Phoenix sylvestris*), sodium and potassium were found to be high. Amino acids such as glutamic acid, serine, aspartic acid and threonine were found to be higher in palm saps of *Cocos nucifera, Borassus aethiopum, Caryota urens* and *Phoenix dactylifera.* The amino acid profile of these saps is presented in Table 2.

Phoenix sylvestris sap was reported to be abundant in proteins, lipids, fibres, and 13 different vitamins (Salvi and Katewa 2012). Similarly, sap from *Borussus flabellifer, Cocos nucifera* and *Phoenix sylvestris* were reported to contain inorganic components like calcium, magnesium, iron, sodium, potassium, zinc, copper and phosphorus. Sap from *Nypa fruticans* showed high amounts of sodium and chlorine content when found near the coastal region, and it is also rich in essential amino acids, minerals and vitamins (Chinnamma et al., 2019; Van Nguyen et al., 2016). It has been noted that the sugar rich sap aids in the growth of microbes and result in the degradation of protein or autolysis of yeast that may decrease the amino acid content of the sap (Hebbar et al., 2018). Moreover, palm sap is rich in inulin, a polymer of fructose, which can be used as a natural sugar replacer and has prebiotic properties. Palm saps are rich in phytonutrients such as flavonoids, anthocyanidin and polyphenols that influence the quality, sensory and antioxidant potential of the drink. Generally fresh sap contains about 0.33 g l^{-1} of total phenolics (Xia 2011). Vitamin C is the major vitamin found in the sap along with vitamin B complex (Hebbar et al., 2015). *Borassus flabellifer* is reported to have vitamin C content of about 13.25 mg 100 ml^{-1} which is higher than *Cocos nucifera.* The pH of the fresh sap is near to neutral (6.6–6.8) and the fermented sap has a pH range of 3.5–4.0. It is also rich in phytochemicals and volatiles such as aromatic hydrocarbons, aliphatic ketones, acids, alcohols, heterocyclic compounds and esters. Among all, acetic acid and 2- butanol were identified as the major volatile compounds (Purnomo 2007).

Table 2. Amino acid profile of different palm saps (100 ml)

Amino acids	*Cocos nucifera*	*Borassus aethiopum*	*Caryota urens*	*Phoenix dactylifera*
Tryptophan	1.27	-	-	-
Lysine	0.32	3.90	1.00	0.04
Histidine	1.19	2.10	6.00	0.03
Arginine	0.35	2.40	10.00	0.18
Aspartic acid	11.22	9.30	8.00	0.86
Threonine	15.36	3.60	1.50	0.05
Serine	8.24	3.90	14.00	0.20
Glutamic acid	34.20	8.90	-	1.18
Proline	3.52	5.00	3.00	0.02
Glycine	0.47	5.00	2.00	0.06
Cystine	2.56	0.10	4.00	-
Alanine	2.11	5.00	4.00	0.21
Valine	0.38	5.00	-	0.04
Isoleucine	0.48	3.60	4.00	0.05
Leucine	0.31	6.10	1.00	0.03
Tyrosine	0.78	2.80	2.00	-
Phenylalanine	-	3.60	2.00	0.07

Volatiles are formed from the presence of abundant polar side chains of amino acids like glutamine and aspargine. Due the catabolism of these amino acids, a large amount of free amino groups is released which reacts with the carbonyl group of the reducing sugar present in the sap through a series of secondary reaction to form the flavouring compounds (Hebbar et al., 2018). The unique flavour of the sap is due to the presence of these compounds. Borse et al., (2006) investigated the volatile compounds present in the sap and confirmed the presence of compounds like 1- hexanol, ethyl lactate, tetradecanone, phenyl ethyl alcohol, 2- methyl tetrahydrofuran, 3- hydroxy- 2- pentanone. The presence of dodecanoic acid and palmitoleic acid along with ethanol and ethyl esters are responsible for making the sap unpalatable upon fermentation, whereas ethyl lactate, 3- hydroxy- 2- pentanone, and farnesol are commonly present in both fresh and fermented sap. The catabolism of amino acids and sugars converge at pyruvate which acts as precursor for the formation of ethyl lactate and 1-Hexanol (Pighin et al., 2016). The presence of phenylalanine in the sap is converted to phenylacetaldehyde which upon enzymatic conversion by phenylacetaldehyde reductase yield phenyl ethanol (Sheng et al., 2018). The acetyl-coA provides acetyl group to the citric acid cycle which is converted to malanoyl coA and finally linoleic acid which by the action of lipoxygenases leading to the formation of 2-hydroxy-3-pentanone and 3-hydroxy-2-pentanone (Vichi 2010).

Due to the fermentation process, the pH of the palm sap decreases, when it goes below pH 7.0 amadori rearrangement takes place especially 1,2 enoisation resulting in the formation of furfural (Naknean 2010). The furfural undergoes a series of reduction reaction to form 2-methyltetrahydrofuran (Wang et al., 2019). The sap is rich in sucrose which is broken down to monomeric units of glucose and fructose due to microbial action. The fructose is again joined together as a polymer by the enzyme fructan 1-fructosyltransferase resulting in the formation of Inulin (Karimi et al., 2021). These volatile components not only affect the quality attributes of the sap but also the boiling point and heating duration when the sap is subjected for downstream process (Martins et al., 2001). The high sugar content (sucrose, glucose, fructose, maltose, raffinose) makes it susceptible to fermentation by the microflora present in the sap that enter from the environment. The reducing sugar content is considered as indicator of freshness of the sap, and it should be less than 1.0 g 100 ml^{-1} (Hebbar et al., 2015). Fermentation initially results in the formation of lactic acid and finally alcohol was formed. The action of microflora also contributes for the generation of phenolics *viz.,* protocatechuic acid, gallic acid, galangin and caffeic acid that are identified as major phenolics in the fermented sap (Xia 2011).

6. Tapping of Different Palms

6.1. *Arenga* sp

The genus includes 24 species and most of them are utilized locally by some tribal communities as an integral part of their traditional medicinal system. *Arenga pinnata*, *A.obtusifolia*, *A. micrantha,* and *A.wightii* are four members of the Genus *Arenga* reported in India. The tribal communities like *Mudhuvans* of Idukki district, Kerala, India depend on the palm mainly for its starch from the pith and toddy collected by tapping its inflorescence. The inflorescence of the palm is used for religious ceremonies and the leaves are used for thatching the huts and pandals in their settlement. The palm is tapped from the unopened spadix and the sap is collected in a long bamboo from April and is consumed as wine without fermenting. The fresh sap is also given to patients suffering from jaundice. Vinegar and jaggery are the two value added products derived from the palm wine. The *Malampandaram* tribes of Achancovil forest range in the Kollam district of Kerala, India uses the white powdery material from the leaf of the palm for wound healing.

There are significant differences between the methods practiced by the different tribes. Tapping is done on the male inflorescence spadix as the sap from female inflorescence is of inferior quality and the spadix warrants more effort during preparation of tapping. A wooden mallet is used for slowly hitting the trunk beneath and above the stem of the inflorescence before incision. The tapper cleans the male inflorescence stalk from its bracts as soon as the flowers are open. The clean stalk is then slowly beaten for several minutes each day for about 2 or 3 weeks at a distance of about 45-60 cm from the stalk base, until the stalk becomes swollen; later the stalk is cut at the swollen section. The sap is collected in a bamboo pipe. In every tapping, the apex of the stalk is sliced 1 to 5 mm depending on the pre-treatment. Another method consists of splitting the stem of the inflorescence into 2 parts and removing secondary stems except the last four that are clubbed into a bamboo pitcher where the juice will fall. The four secondary stems extremities are cut and then bruised with wood pliers twice a day for 5 days. During the next two days, they are placed in a bamboo container having one litre of water. They are bruised again for the next two days and then the production starts. The simplest method consists of inserting a bamboo pipe into the base of the male flowers and collecting the sap into a gourd.

In some places of Sulawesi (Indonesia), *Arenga pinnata* and *Arenga wightii* are cut down, the leaves are lopped, and the trunk is sliced at intervals.

Tapping Arenga is a very labor-intensive activity, which is done on a regular schedule twice-a-day. Beating and preparation of 20 inflorescence stalks (of which 12-16 will produce each for 4 months) takes around 2.5 h each day. One healthy person can tap 12-16 trees on an 8 h working day or more if the trees are close enough. On average, a person taps about 10 trees and 10 workers are necessary for palms planted in an area of one hectare. Tapping must be done daily or otherwise the sap flow will diminish rapidly as process of tissue healing occurs. To slow down sap fermentation, the container is sterilized with the smoke of a wood fire.

6.2. *Borassus flabellifer*

The traditional approach involves tapping the top shoots and collecting the dripping juice in hanging earthen pots (in some regions a plastic or bamboo bottle). The juice collected in evening or after fermentation becomes sour, and is called tadi in Marathi, hta-yay in Myanmar and Bhojpuri. Tapping is done by bruising the stalk with a wooden instrument to crush the embryo flower or fruit within and removing a thin slice at the end for several days. On eighth day, the sap begins to exude, which is collected in an earthen vessel attached to the stalk. The vessel is smeared with lime (chunam) inside to prevent the fermentation of juice. Climbing of the tree is done during morning and evening to cut a small thickness of the stalk. However, the sap, which has collected since the last ascent, is usually brought down only in the morning.

The quantity of juice from a male tree is only two-thirds of that obtained from a female tree, but is said to be sweeter. Much of the unfermented sap is consumed just as it is collected from the tree, but the supply being more than the demand in this form, a major portion is boiled to produce jaggery or converted into sugar candy.

In Sri Lanka, four different methods are used according to the sex and age of the inflorescence.

1. Two-week old male inflorescences are cleaned from their outer sheaths and left to dry for 3 days and later a fresh surface is cut every day and about 1 l sap is obtained daily for 45 days.
2. All the spikes (12) of one-month old male inflorescences are pressed, stroked and grouped 3 to 6 together to be fitted to a pot for tapping and collecting the sap.
3. The main axis of young female inflorescences is struck with an iron rod to soften the tissues, and a fork is used to crush the nodes where fruits would arise.
4. Two to three-month old female inflorescences, which already bear fruits, are sliced daily for collecting the sap.

Tapping techniques of *B. flabellifer* are similar to that of coconut except for slight differences. In Cambodia, if a tree produces many inflorescences at the same time, some of them (1 to 7 according to tree, climate and location) are left as such while the others are sliced and crushed for 5-8 days and then preserved (3 to 5 months) for juice collection some months later. Different tongs are used according to the sex of inflorescence and stage of tapping. Clay pots, bamboos internodes, calabashes or leaf-buckets are used for sap collection. In Central Myanmar, sap collection can be continued all year round but with low yields between November and January. In the upper Myanmar, the 8-month tapping season consists of three tapping phases and four different tapping techniques are adopted depending on the sex of the palm.

6.3. *Caryota urens*

Caryota urens is commonly known as Fishtail palm or Indian Sago and locally as Sulphi in Bastar plateau agro climatic zone of Chhattisgarh, India. The fresh sap from this tree is very nutritious and is called as 'Bastar Beer' in India.

The art of tapping a *C.urens* flower is very much a village craft requiring a high degree of traditional expertise. Palms aged 20-30 years produce the most sap and they are tapped every day morning and evening from October through mid-June. When the inflorescence emerges sufficiently (about 90 cm) just prior to unfolding, it is supported by a long pole (120 to 150 cm) forked at the top and manoeuvred into a horizontal position for the convenience of tapping. The immature inflorescence has a cone shape, while mature inflorescence has an inverted crescent shape. The inflorescence by this time has reached its maximum length and is on the verge of unfolding. Such mature inflorescence is subjected to a process of stimulation, which arrests the growth, but does not prevent it.

Initially, the enclosing bracts are removed and the peduncle is bruised gently by tapping with a small stone or the handle of the tapper's knife. A paste of a ground mixture of *Pogostemon heyneanus* ('kollan kola') leaves, slaked lime, lamp black and other additives such as salt, garlic, mustard, lime and young leaves of *Toddalia asiatica* ('Kudumirissa') and *Nepenthes distillatoria* ('Bandura') is applied to the bruised area. The treated inflorescence is bound along its length with cord made from woody climbers and left exposed for a couple of days. If the weather is dry, the inflorescence is shaded using a bract from the areca nut palm *(Areca catechu),* leaves of *Agrostistachys hookeri* ('Mahabeeru') or *Coscinium fenestratum* ('Weniwel). Of late, polythene is increasingly being used. The main inflorescence axis yields the greatest amount of sap, and hence all lateral branches and spikes are removed before trimming the main axis with a very sharp knife. The exuding sap is collected in a vessel, and this process carried out twice a day for a period of about two months.

6.4. *Cocos nucifera*

The development of female flowers inside the spathe causes a swelling at the basis of the spadix, which is considered as an acceptable stage for tapping. Such selected inflorescence is first trained by a gentle uniform beating all over the spadix twice a day, in the morning and in the evening. It is also tied all around with a strong coir or coconut fibre string to prevent it from bursting. After a week, about 7-10 cm of the tip of the spadix is cut off. On the third day, the cut end is carefully pounded. Beating is continued for another week, while an earthen pot (in Sri Lanka) or a bamboo tube (Thailand) is hung from the tip of the spadix to collect the oozing liquid from the cut surface. The free end of the spadix is gradually bent down. The sap generally begins to drip after 12-15 days up to 35 days after beginning of the tapping process according to the tapper's skill, seasonal conditions and health of the palm. Slicing of about 2 mm of spadix and collecting sap are done twice a day while tapping and bruising operations are discontinued. However, light tapping of the end of the spadix with a mallet may be continued and the sliced surface of the spathe is sometimes smeared with a mixture of bruised leaves which contains saponin to stimulate the sap flow. If the trees are tapped for sweet juice, they are collected in the morning. A single spadix can be tapped until it is reduced to a stump about 10-15 cm in length. About three weeks before reaching this point, another spadix is prepared to ensure continuity of sap production without any break.

6.5. *Corypha elata* (Syn.: *C. utan*)

In this palm, the entire growing point of the terminal inflorescence is excised just before its development starts. A shallow depression is scooped out at the top of the naked stem, and the juice is collected from the depression (Kovoor 1983; Redhead 1989). In Roti (Indonesia), the technique consists in breaching the inflorescence (Fox 1977). Incisions of the flower stalk are also reported (Pethiyagoda 1978) for producing sap.

6.6. *Elaeis guineensis*

Destructive tapping is the most common method followed in West Africa and particularly in Ivory Coast and Ghana. Just before the fructification, the young tree is felled down by loosening the roots with a spade and pulling down the tree. For older trees, the trunk is cut at 30-60 cm. All fronds in the meristematic region are removed after one or two weeks. A cavity is made in the terminal bud, which is either a rectangular well about 7.5 cm deep with a round hole bored in its floor or a bowl-shaped hollow 10 cm x 9 cm, about two-third (15 cm) the diameter of the stem in depth. A brand is put inside to activate the sap flow and reduce the infestation of the well by insects and other microorganisms. Collection of the sap using a bamboo gutter (15 cm long) which goes into a container (calabash) under the stem commences 2 to 4 days later. Collection is done twice a day and each time, the tapper cuts 1 mm layer of tissue from the terminal bud and flares with fire the stem surface around the tapping wound. The sap flow lasts for 30 to 60 days.

A less destructive method is practised in Nigeria (Igbos) which consists of piercing the stem of the standing tree up to the bud (Kovoor 1983). Several leaves are removed from the side of the stem. A rectangular hole (8 cm x 5-10 cm) is cut in the stem to a depth of about 30 cm or more according to the size of the palm tree. A suction-pipe made of bamboo is used to collect the sap into a jug hung on the stem. As the sap flow is very intense, the tapper must climb the tree 3 times a day to collect it. Each time, he cuts away a thin layer of tissue from the wound surface in a downward direction. The length of the rectangular hole can thus extend to 30 cm or more according to the duration of sap flow. The tree dies 2 or 3 weeks after being tapped.

A non-destructive method consists in tapping of the male inflorescences. It is especially practised by the Jorubas in Nigeria, but it is also found in Benin and Ivory Coast. All fronds below an unopened male inflorescence (before the differentiation of the flower spikes starts) are cut off and the leaf subtending it is removed to obtain access to the inflorescence enclosed in its spathes. About 7.5 cm above the base of the male inflorescence, a neat triangular hole of about 2.5 to 6 cm deep in the main stem of the spadix and 2.5 cm on the sides is made in the front spathe. The hole is then covered and the following day a funnel of small bamboo (16 cm long and 2 cm in diameter) is inserted in the cut in such a position that the sap flows into calabashes (vessels). To ensure continuous flow, the hole is extended downwards by removing a daily slice of about 3 mm. The tapping is stopped when the base of the male inflorescence is reached. The sap is collected during morning and evening and a new slice is taken at each collection. Palm trees tapped in this manner set fruit one to two years after tapping.

6.7. *Hyphaene coriacea*

Palm clumps are selected and large stems within these clumps are prepared for tapping after removing the undergrowth and leaf spines. The selected stems are trimmed using a razor-sharp bush knife to initiate sap flow. The stem and young leaf bases are cut at a slight angle to guide sap onto a leaf gutter and into the clay collection container. By the end of the tapping period, these young leaf bases are removed by trimming 2 to 3 times in a day for a period of 5- 7 weeks. Therefore, the terminal bud is cut causing death of the stem, but the clump survives (Cunningham 1990).

6.8. *Hyphaene thebaica*

As *Hyphaene thebaica* is a multi-stemmed palm, tapping of sap may not result in the death of the clump. There is also a less destructive technique. The selected stem is cleaned off all leaves followed by removing the bases of the future leaves leaving the terminal bud as such. New cuts are practised up to 4 times a day for a couple of weeks and are generally stopped before reaching the terminal bud. Trained tappers will reach this terminal bud to extract more sap before having to prepare another palm for sap extraction.

6.9. *Nypa fruticans* (Nipa palm)

Nypa fruticans, known as the 'mangrove palm' or '*golpata'* in both Bangladesh and India, is widely grown throughout Southeast Asia however, mostly at the subsistence level. Nipa palm is also much valued for the sweet sap tapped from the stalk of the inflorescence. The tapping normally begins from the shoots after four years of growth and continues up to 15 years or more. The shoots from 9-12 years old stems are the high yielders, providing up to 1.5-1.9 l of sap stem^{-1} season^{-1}. The sap yield from stems of 15 years or more will be low. Tapping commences from the first week of December when the fruit begins to mature and turns tan, and continued until the next mid-March. The stalk of inflorescence is pulled down and every morning and evening, a thin slice of stalk is removed, and sap collected in a container. The tapping life depends on the length of the stalk which can vary from 0.6 to 1.0 m in the Philippines, 1.4 m in Sumatra and 1.9 m in Papua New Guinea

(Hamilton and Murphy 1988). To slow down the fermentation of the sap, the bamboo recipients (2 l capacity) are washed with lime and water.

In Papua New Guinea, a sophisticated protocol was developed by Päivöke (1985). The pre-treatment pattern was bending the stalk 12 times in one direction, patting the length of the stalk back and forth with the bare hands 64 times and kicking the base of the stalk 4 times, which is repeated for 4 times a week. Yields of 1,800 ml day^{-1} $tree^{-1}$ over a month were obtained in intensively treated younger palms whereas, only 155 ml day^{-1} $tree^{-1}$ were produced when the treatment was done only once a week. The duration of treatment depended on the state of development of the flower stalk.

6.10. *Phoenix sylvestris* (Wild Date Palm)

P. sylvestris is found in drier-to-moist tropical and subtropical climatic zones. The palm grows naturally and is cultivated around homesteads, periphery of farmlands and in marginal lands along the roadside and canals, even on fallow land. It can survive in disturbed areas, such as wastelands or seasonally inundated areas. *P. sylvestris* is widely distributed in India, Pakistan, Myanmar, Nepal, Bhutan, Bangladesh, Mauritius, China and Sri Lanka.

For tapping, the stem is pierced up to the terminal bud (Kovoor 1983). In Bangladesh, the oldest leaves are removed on one side of the palm tree along about 35 cm. The bases of the petioles and the sheaths are carefully removed. A triangle piece of bark pointing downwards is cut carefully so as not to expose the sap-supplying inner zone. A week later, as the fine covering of soft tissues gets hardened and begins to crack, this covering is removed without damaging the inner zone. After resting for 12 to 14 days, a lateral incision is done in the triangle and the sap exuding from the scooped surface is run through a bamboo spout and collected in a vessel. A new incision is made after two days and then there is a 4-day period of rest, and the same cycle is repeated until the heart is reached. The following year, the same treatment is done on the other side of the tree which gives a zigzag configuration to the trunk. The tapping can be practised every year for several decades (average of 25 years). To slow down the fermentation of the sap, the earthen vessel is coated with lime and in Bangladesh they are well smoked every morning.

Tapping the wild date palm in India is a very well-developed cottage industry. However, the palms don't offer an alternative product because its fruits are not attractive for human consumption. Though palm tapping is done annually, the whole crown of leaves is not removed, thus leaving a great part of the productive capacity of the palm (Barreveld 1993).

6.11. *Phoenix dactylifera* (Date Palm)

In this palm, the entire growing point is cut off and a shallow depression is scooped out at the top of the naked stem, through which the sap is collected. Another technique is cutting off the fronds and exudation of sap occurs in the cavity created by defoliation. Tapping a date palm is a severe intervention as it deprives the palm of most of its leaves and food reserves (Barreveld 1993). Therefore, great care is needed during the scarring which is performed twice in a day, because if it is continued too far, the palm will die. The cone, canal and spout are to be re-adjusted every 20 days or so. Once tapping is completed in three or four months, three to four years are needed for the tree to bear a full crop of fruit again and thereafter another five years for the next tapping.

6.12. *Phoenix reclinate*

In Côte d'Ivoire, the tree is uprooted, all leaves are removed, and the terminal bud is cut. A cavity is made in which a brand is put in order to activate the sap flow. Collection of the sap starts 3 to 4 days later through a gutter linked to a calabash. Collection is done twice a day, each time, the terminal bud is further cut. It lasts for 1 to 2 months. Another technique consists in incising the lateral part of the young stem. In south-eastern Africa, the same technique as described by Cunningham (1990) for *Hyphaene coriacea* is being used.

6.13. *Raphia* sp

Raffia palm (*Raphia* sp) is among the most diverse and geographically wide spread. There are over 20 species of raffia palm which originated from West Africa. When several species are available, tapping is based on accessibility, distance, abundance, availability, and amount of sap produced per unit of time. Morphology also plays an important role in choosing a right species for tapping. *R. hookeri* is preferred in the littoral and central regions of Cameroon and is easily tapped since it has single stem, while *R. vinifera* is preferred in the Western Highlands. If not much species diversity is observed the available species is tapped even though in accessible (e.g., *R. zamiana* in South Cameroon). In Madagascar *R. farinifera* is used for wine extraction (Kamga et al., 2020).

The sap of Raphia palm is drunk by millions of people in Africa as a source of sugary beverage (palm wine). The sap ferments within few hours and turns to an alcoholic drink referred to as palm wine in Nigeria (Bassir 1962). The drinking of palm wine signals the start and end of all social activities undertaken by 50 million people of southern Nigeria. The Raphia palms alone contribute about 20% of the palm wine drunk in its unfermented as well as fermented states (Obahiagbon and Osagie 2007). To obtain the sap, the Raphia palm is tapped daily in the traditional manner described by Tuley (1964). There are two main methods for sap extraction: down wine tapping and inflorescence tapping (Tuley 1995). In the former, individual trees are felled and a container is placed under the cut inflorescence to collect the sap. This is destructive and practiced in *R. hookeri* and *R. monbuttorum*. In the latter method, tappers climb to the base of the crown and apply an incision near the emerging inflorescence primordium allowing the sap to flow out. This method necessitates expertise and skill of the tapper to climb the palm.

Some native tappers in the south of Nigeria apply heat to stimulate the flow of the sap at the panel where the receiver is hung (Otedoh 1990). The sap is colourless and very sugary/sweet (Obahiagbon et al., 2007). The sap can also be collected from the upper stem pierced to death as it is currently practised in West Africa; the method is the same as the one described for oil palms by Igbos in Nigeria. The tappers extract the sap just before the appearance of the spadix and tapping is done on flowering trunks.

7. Importance of Palm Sap

Inflorescence saps of various palms from different geographical regions around the world command a great importance. Ignorance and lack of awareness of this nutritive drink has made the sap unpopular in urban areas (Chandran and Francis 2016). Sugarcane juice and palm sap are comparable as both are rich in sugars and other phytochemicals, however, palm sap is superior due to its high nutritional value. Palm sap has been proven to have high amount of calcium (0.149 g 100 g^{-1}), phosphorus (0.011 g 100 g^{-1}) and iron (0.2 g 100 g^{-1}) which are absent in sugarcane juice (Perdrix et al., 1999). Coconut palms also produce tender water which is widely consumed all over the world due to its health benefits and profound awareness (Naik et al., 2020).

However, palm sap is also of same origin and is often neglected as people are unaware of its therapeutic and medicinal benefits. The sap has more calorie than milk and proven to be better than mineral water and has less calorific value (Luis et al., 2012). With all the nutraceutical and functional benefits, the value-added products produced from this sap also have income generating potential. The sap was compared with a soft drink named 'Colonche' consumed by the inhabitants from the arid places of northwest Mexico and palm sap was reported to have medicinal properties as it contains more amount of vitamin C than the soft drink. In a comparative study by Asghar et al., (2020), coconut sap and sugar palm juice were found to have higher antioxidant and nutritional value than sugarcane juice. Higher amount of fructose, glucose and minerals (sodium, potassium and iron) was also found in coconut sap as compared to sugar palm and sugarcane juice. Being rich in medicinal properties and therapeutic potential, this sap is often converted to value added products that are of great commercial importance. Tapping of this refreshing sap may have potential benefits of generating income for farmers and agro based industries. To achieve the penitential benefits, it is important to identify the factors like microeconomic, political issues and the link to markets that affect the smallholders involved in palm sap harvest. Increasing the production and extending transportation to urban areas from sub-urban regions would be a key factor for higher benefits.

8. Multipurpose Utilities and Sustainable Integrated Production Systems

Most palm trees have multipurpose uses. Nevertheless, they are not always compatible. Sap production is at its maximum just before or during fruit formation. Tapping the tree competes with the production of the ripening fruit. Tapping can also stimulate fruit production: a young coconut palm tapped during 6-12 months for sugar production will then produce more nuts. A technique called sequential coconut toddy and nut production has been developed in the Philippines at the Davao Research Centre. The first half of the spadix is tapped, and the second half is left for fruit production as female flowers that develop to mature nuts are situated in this lower portion. Nut and copra yields are about 50% lower than non-tapped palms; however, this technique has been demonstrated to be very feasible and highly profitable for small producers.

Arenga pinnata can be tapped when they are between 12-15 years and more than 30 years old. The palms can be cut for sago production (Sumadi 1988). In West Java, sago is obtained from trees aged 10–12 years, with no prior tapping, as farmers believe it would reduce the starch content in the trunk (Mogea et al., 1991). In Eastern Nigeria, oil palms abandoned as uneconomical for bunch production often provide good economic returns for wine production before old plantings are cleared and replanted (Tuley, 1965). Various palm-crop associations exist in Bangladesh. *Phoenix sylvestris* and *Borassus flabellifer* are commonly intercropped with rice, wheat, chickpea, mustard, jute, lentil, potato, linseed, winter vegetables, and sugar cane. Palm trees offer sustainability advantages compared to other crops. In parts of West Java, where *Arenga pinnata* is grown in groves, soils are more stable and productive than areas cultivated with cassava. This tree exhibits great ecological tolerance, thriving on unproductive, erosion-prone sites like steep, dry slopes, such as coffee orchards in North Sulawesi. In Cambodia and India, *Borassus flabellifer* is often planted along paddy field boundaries. Its small crowns and the wide spacing between trees (10–15 m) minimize shading effects on understory crops. Similar to *Arenga pinnata*, it flourishes in infertile, arid regions and withstands prolonged flooding. Highly resistant to pests and diseases, it requires minimal cultivation. With a deep root system, it helps prevent erosion and stabilizes dunes in sandy plains. Like *Corypha elata*, it is a fire-resistant pioneer species, commonly found in areas affected by slash-and-burn agriculture. In Myanmar, it serves as a windbreak in groundnut fields. On Savu and Roti islands in Indonesia, where soil fertility is a major challenge, *Borassus* leaves are used in semi-permanent gardening practices, replacing traditional slash-and-burn methods in Timor and Sumba. Burning old leaves in fields helps maintain fertile gardens near homes. *Borassus* forests display a unique nutrient cycling pattern, supporting stable agriculture and aiding in the recovery of disturbed sites. In the Peruvian Amazon, *Mauritia flexuosa* forms dense populations in seasonal swamp forests, growing on waterlogged or sandy soils typically unsuitable for agriculture. Meanwhile, *Nypa fruticans*, unlike sugar cane, does not compete with other crops for agricultural land - except in cases of full-scale reclamation of mangrove areas.

9. Value Added Products from Palm Saps

9.1. Palm Sugar

Value-added products developed from the sap or nectar of palms are used as traditional and alternative sweeteners in drinks and food products across India, Indonesia, the Philippines, and Thailand. In Malaysia, only coconut sap is utilized for value addition. Indonesia and the Philippines are the world's largest producers of palm sugar, which is prepared through the caramelization of palm sap (Saputro et al., 2019). Beyond providing sweetness, palm sugar influences the flavor, texture, and color of food products (Jose et al., 2018). When mixed with corn sugar, palm sugar has been shown to slow digestion, thereby lowering the glycemic index (GI) compared to refined cane sugar (Jayanudin et al., 2019). Palm sap is also rich in essential nutrients, including amino acids, vitamins (B1, B2, B3, and B6), antioxidants (such as 2,3-dihydro-3,5-dihydroxy-6-methyl-4(H)-pyran-4-one), and dietary fiber (inulin), all of which offer various health benefits (Swamy, 2013).

9.2. Palm Wine (Toddy)

The natural microflora present in palm sap ferments the sweet sap into a milky white liquid known as palm wine, with an alcohol content of 4-4.5% (Lasekan and Abbas, 2010). This alcoholic drink, derived from members of the Palmae family, is commonly referred to as 'toddy' and is popular in the tropical regions of Asia, Africa, and South America. In India, toddy is fermented from the sap of palmyra palm, silver date palm, and coconut palm (Chandrasekhar et al., 2012). In Nigeria and parts of Africa, it is traditionally used in rituals to ward off negative supernatural forces (Chandrasekhar et al., 2012). The geography and species of palm tree influence the microbial diversity of the fermented sap, offering the potential to produce a wide range of secondary metabolites that contribute to the distinctive sensory attributes of toddy (Karamoko et al., 2016). While palm wine contains alcohol and can act as a natural disinfectant, its high sugar content makes it unsuitable for clinical or household disinfection purposes (Perdrix et al., 1999). Studies have shown that palm wine exhibits antibacterial activity against *Staphylococcus aureus* and *Pseudomonas aeruginosa*. Moreover, it possesses therapeutic benefits due to the presence of the antipsychotic agent gevotroline and the anti-carcinogenic agent mitoxantrone (Djeni et al., 2020).

9.3. Palm Syrup and Honey

The supersaturated form of palm syrup results in a highly viscous solution resembling honey. It is used in Ayurveda for fitness, well-being, and as a natural sweetener in food products (Luis et al., 2012). Palm syrup is extensively produced in the Canary Islands and South America from the sap of *Phoenix canariensis*. The processed sap is known as 'miel de palma,' while a similar product, Kithul syrup, is made in Sri Lanka (Francisco-Ortega and Zona, 2013). A comparable syrup is also produced from *Jubaea chilensis* (Chilean wine palm) in Chile - an endangered species - reported to have higher levels of vitamins, potassium, sodium, and antioxidants compared to other palm-based syrups (Van Nguyen et al., 2016). When stored properly, palm syrup can last for several years without the need for preservatives (Asghar et al., 2020). Recently, its unique flavor has boosted its popularity in the confectionery industry and among consumers (Misra, 2016). Similar to palm sap, palm honey has been shown to increase serum antioxidant capacity, reduce lipid peroxidation, and help prevent atherosclerosis (Hegazi and El-Hady, 2009). It is notably rich in myo-inositol (3–8 mg/g), a compound used to manage diabetic neuropathy, depression, Alzheimer's disease, panic disorder, and polycystic ovary syndrome (PCOS) (Luis et al., 2012). Dark palm honey from Egypt exhibits higher antioxidant activity compared to acacia and coriander honey. Its chemical characterization reveals the presence of compounds such as succinic acid, 3,4-dimethoxycinnamic acid, 2,5-dimethoxycinnamic acid, picolinic acid, 4H-pyran-4-one-5-hydroxy-2-hydroxymethyl isomer, and pyridine carboxylic acid (Hegazi and El-Hady, 2009).

9.4. Palm Jaggery

Palm jaggery is produced by boiling and evaporation of sap in small quantities. It is commonly known as '*gur,*' '*karupatti,*' '*vellam*' or '*pana vellam,*' '*thaati bella*; and '*olebella*' in India. In Sri Lanka, jaggery (hakuru) is produced from the sap of *Caryota urens* (Dalibard 1999). This clear brown semi solid crystalline structure is obtained by caramelization and has a pH between 6.5

and 7.0 (Nighitha and Sudagar 2017). It is widely used as a sweetening agent in the traditional beverage industry, and is available in solid, liquid, powder as well as granular forms. It has many medicinal properties and contains minerals like calcium, phosphorus and iron (Vengaiah et al., 2013). Further, it is low in sodium and high in vitamin B12, vitamin C and potassium that lower the risk of hypertension, anaemia and oedema (Vengaiah et al., 2017).

9.5. Palm Sap Vinegar

The palm vinegar has various uses such as medicine, food additive and preservative and other culinary uses (Samuel et al., 2016). It is extensively consumed in many countries of Asia and Pacific Islands. In the Philippines, saps from *Arenga pinnata, Cocos nucifera* and *Nypa fruticans* are used for making vinegar (Van Nguyen et al., 2016). It is produced from toddy (Ghosh et al., 2012). Alcohol fermentation is the first step followed by acetic acid fermentation. The acetic acid, thus, produced during fermentation is responsible for the flavour and odour of vinegar. Natural palm sap vinegar has better export potential than synthetic vinegar (Samuel et al., 2016). The aqueous extract of palm vinegar showed hypoglycaemic effect when tested in male Sprague-Dawley rats (Yusoff et al., 2015).

9.6. Molasses

Palm molasses is a viscous produce obtained from the refinement of palm sugar. Its quality varies according to the age of the tree, method of extraction and amount of sugar (Widyaningrum et al., 2019). It is mostly produced ((i) in rural areas and households by collecting and boiling the sap, or (ii) in pilot scale level (iii) during storage of bagged and humid dates (Hebbar et al., 2018). Palm molasses from date palm of Oman was used as bio-surfactant. It was also reported to have better stability under varied temperature, pH and salt concentration (Al-Bahry et al., 2013). Various biochemicals like bioethanol, acetic acid and citric acid can be generated by the fermentation of molasses (Misra 2016).

9.7. Levan Production

Palm sap is a good substrate to produce microbial levan polysaccharide. The potential usage of fresh coconut inflorescence sap as a fermentation medium for levan production using *Bacillus subtilis* was exploited and it was found that levan yield is highly dependent on fermentation conditions such as temperature, pH, and agitation speed. Hence, palm sap can be used as low-cost source for levan polysaccharide production and can be produced in less time. Since the palm sap is of organic in nature, the biochemicals present in the sap serve as growth enhancers and act as a suitable fermentation medium without addition of chemical constituents (Mummaleti et al., 2021a, Mummaleti et al., 2021b).

9.8. Other Value-Added Products

Various value-added products *viz;* candy, chocolate, cookies, soft drinks, squash etc; are manufactured from palm sap. Products like chocolates are low in calorie making it safe for ingestion for diabetics (Vengaiah et al., 2017). Confectionery items like cookies made from the sap are also suitable as snacks for diabetics upon complementation with multigrain, oats, whole wheat etc. (Karseno et al., 2018). In Cambodia and Indonesia, fresh sap is used as an essential ingredient in pig diet (Dalibard 1999; Rodríguez and Preston 1997). Drinks made from the sap are beneficial for health as it is rich in minerals and vitamins (Misra 2016). A concentrate (200 ml) prepared from 1000 ml sap can be diluted as a drink and used as a bread spread (Chinnamma et al., 2019).

10. The Way Forward for the Sustainable Utilization of Sap-Yielding Palms

Approximately 2,800 species of palm trees are grown in tropical, subtropical, and no-frost temperate zones around the world. Palms are cultivated commercially for food production and to create aesthetically pleasing landscapes. Palm trees are believed to be the oldest fruit-bearing trees on Earth. Except for a few species, most palms grow naturally and are sporadically spread across regions. They are considered hardy species with known climate-resilient traits.

Many palms, including domesticated species like coconut and date palm, are primarily used for tapping sap (*neera/Kalparasa®*), a highly nutritious and healthy drink that is further processed into various value-added products. However, the biggest challenge is sap collection from the tree. It can be extracted from the unopened spadix (a non-destructive method) or directly from the stem near the crown (a destructive method). In both cases, the sap drips slowly, and during the collection process, it often undergoes microbial or enzymatic fermentation, turning it into an alcoholic beverage. Sap yield varies greatly across species, ranging from 1 to 80 litres per day. Tapping and collecting sap requires climbing the tree, which poses a significant challenge. Therefore, breeding approaches are required to identify and develop dwarf varieties that are convenient for tapping. Additionally, optimizing in vitro multiplication protocols to produce seedlings from high sap-yielding trees is crucial.

Further studies are also warranted to assess the wider adaptability of these species under extreme climatic conditions to ensure sustainable production. Integrating palms into agroforestry systems could enhance biodiversity, improve soil health, and provide additional income sources for smallholder farmers in tropical regions. Moreover, improving market access for small-scale farmers by promoting fair trade practices, strengthening supply chains, and connecting local producers to global markets for palm-based products would be beneficial. Finally, engaging local communities in the management of palm resources is essential to ensure that the benefits of palm production are equitably shared, particularly by empowering women and indigenous groups who traditionally rely on palms for their livelihoods.

Conclusion

In conclusion, palm trees represent a valuable resource for both ecological sustainability and economic development, particularly in tropical and subtropical regions. Their resilience to climate stress, nutritional sap, and multipurpose utility make them ideal candidates for integrated farming systems. However, to fully harness their potential, targeted efforts are needed - ranging from the development of dwarf and high-yielding varieties to improved sap collection methods and propagation techniques. Supporting smallholder farmers through better market access, inclusive community participation, and capacity building will be key to ensuring that palm cultivation not only thrives but also contributes to equitable and sustainable rural livelihoods.

References

Al-Alawi, R.A., Al-Mashiqri, J.H., Al-Nadabi, J.S.M., Al-Shihi, B.I., Baqi, Y. (2017). Date palm tree (*Phoenix dactylifera* L.): Natural products and therapeutic options. *Frontiers Plant Sci* 23(8): 845.

Al-Bahry, S.N., Al-Wahaibi, Y.M., Elshafie, A.E., Al-Bemani, A.S., Joshi, S.J., Al-Makhmari, H.S., Al-Sulaimani, H.S. (2013). Biosurfactant production by *Bacillus subtilis* B20 using date molasses and its possible application in enhanced oil recovery. *Int Biodeterioration Biodegradation* 81*:*141–146.

Asghar, M.T., Yusof, Y.A., Noriznan, M., Yaacob, M.E., Mohd Ghazali, H., Chang, L.S., Manaf, Y.N. (2020). Effect of processing method on vitamin profile, antioxidant properties and total phenolic content of coconut (*Cocos nucifera* L.) sugar syrup. *Int J Food SciTechnol* 55(7): 2762–2770.

Ayernor, G.K.S., Mathews, J.S. (1971) The sap of the palm *Elaeis guineensis* Jacq as raw material for alcoholic fermentation in Ghana. *Tropical Sci*13:71-83.

Babitseng, T.M., Teketay, D. (2013). Impact of wine tapping on the population structure and regeneration of *Hyphaene petersiana* Klotzsch ex Mart. In northern Botswana. *Ethnobot Res Applications* 11: 9–28.

Barreveld, W. H. (1993). Date palms products. *FAO Agricultural Services Bulletin No 101*. FAO, Rome. pp. 262.

Bassir, O. (1962). Observation on the fermentation of palm wine. *West African J Biol Chem* 6: 21-25.

Bernal, R., Galeano, G., García, N., Olivares, I.L., Cocom'a, C. (2010). Uses and commercial prospects for the wine palm, *Attalea butyracea*, in Colombia. *Ethnobotany Res Applications* 8:255–268.

Borse, B.B., Rao, L.J.M., Ramalakshmi, K., Raghavan, B. (2006). Chemical composition of volatiles from coconut sap (*neera*) and effect of processing. *Food Chem* 101(3): 877–880.

Chandran, D., Francis, D.P. (2016). Awareness, perception and satisfaction towards neera health drink: Consumers perspective. *Bonfring Int J Industrial Eng Manage Sci* 6(3): 103–107.

Chandrasekhar, K., Sreevani, S., Seshapani, P., Pramodhakumari, J. (2012). A review on palm wine. *Int J Res Biol Sci* 2(1):33–38.

Chayanika, S., Mummaleti, G., Sivanandham, V., Kalakandan, S., Rawson. A., Anandharaj, A. (2021). Anthology of palm sap: The global status, nutritional composition, health benefits & value added products. *Trends in Food Sci Technol*119: 530-549.

Chinnamma, M., Bhasker, S., Binitha Hari, M., Sreekumar, D., Madhav, H. (2019). Coconut neera—a vital health beverage from coconut palms: Harvesting, processing and quality analysis. *Beverages* 5(1): 22.

Corbishley, D.A., Miller, W. (1984). Tapioca, arrowroot, and sago starches: Production. Starch: *Chem Technol* 2nd Ed, pp 469–478.

Cunningham, A.B. (1990). Income, sap yield and effects of sap tapping on palms in south-eastern Africa. *South African J Bot* 56(2):137-144.

Dalibard, C. (1999). Overall view on the tradition of tapping palm trees and prospects for animal production. *Livestock Res Rural Develop* 11(1):36–77.

Decary, R. (1964). Les utilisations des palmiers `a Madagascar. *J d' Agriculture Tropicale et de Botanique Appliquee*11(8–9):259–266.

Delgado, C., Couturier, G., Mejia, K. (2007). *Mauritia flexuosa* (Arecaceae: Calamoideae), an Amazonian palm with cultivation purposes in Peru. *Fruits* 62(3):157–169.

Dissanayake, B.W. (1977). Use of *Caryota urens* in Sri Lanka. In: Papers of the first international Sago Symposium: The equatorial swamp as a natural resource, Kuala Lumpur pp 84-90.

Djeni, T.N., Kouame, K.H., Ake, F.D.M., Amoikon, L.S.T., Dje, M. K., Jeyaram, K. (2020). Microbial diversity and metabolite profiles of palm wine produced from three different palm tree species in C^ote d'Ivoire. *Scientific Reports* 10(1):1–12.

Dowe, J.L., Ebooks Corporation, ProQuest. (2010). *Australian palms: Biogeography, Ecology and Systematics* (p. 340). CSIRO Publishing.

El-Beltagi, H.S., Mohamed, H.I., Yousef, H.N., Fawzi, E.M. (2018). Biological activities of the doumpalm (*Hyphaene thebaica* L.) extract and its bioactive components. *InTech*. doi: 10.5772/intechopen.74772.

Flores-Gallegos, A.C., V'azquez-Vuelvas, O.F., L'opez-L'opez, L.L., Sainz-Galindo, A., Ascacio-Valdes, J.A., Aguilar, C.N. (2019). Tuba, a fermented and refreshing beverage from coconut palm sap. In Grumezescu AM, Holban MA (eds.), *Non-alcoholic beverages* Sawston: Wood head Publishing. pp. 163–184.

Fox, J.F. (1977). Harvest of the Palm, Ecological Change in Eastern Indonesia. Harvard University Press, Cambridge, Massachusetts, and London, England. 290p.

Francisco-Ortega, J., and Zona, S (2013). Sweet sap from palms, a source of beverages, alcohol, vinegar, syrup, and sugar. *Vieraea* 41:91–113.

Ghosh, D.K., Bandyopadhyay, A., Das, S., Hebbar, K.B., Biswas, B. (2018). Coconut sap (neera) - untapped opportunity of spinoff gains in West Bengal, India. *Int JCurrent MicrobiolApplied Sci* 7:1883–1897.

Ghosh, S., Chakraborty, R., Chatterjee, G., Raychaudhuri, U. (2012). Study on fermentation conditions of palm juice vinegar by Response Surface Methodology and development of a kinetic model. *Brazilian J of Chem Eng* 29(3):461–472.

Gonzalez, L.A., Bustamante, R.O., Navarro, C.R.M., Herrera, M., Toral Ibañez, M. (2009). Ecology and management of the Chilean palm (*Jubaea chilensis*): history, current situation and perspectives. *Palms* 53(2) 68-74.

Hamilton, L.S. and Murphy, D.H.(1988). Use and Management of Nipa Palm (*Nypafruticans*, Arecaceae): a Review. *Econ Bot* 42(2): 206-213.

Hebbar, K.B., Arivalagan M., Manikantan M.R., Mathew A.C., Thamban C., Thomas G.V., Chowdappa P. (2015). Coconut inflorescence sap and its value addition as sugar - collection techniques, yield, properties and market perspective. *Current Sci* 109(8):1411–1417.

Hebbar, K.B., Pandiselvam, R., Manikantan, M.R., Arivalagan, M., Beegum, S.P.P, Chowdappa, P. (2018). Palm sap—quality profiles, fermentation chemistry, and preservation methods. *Sugar Tech* 20(6):621–634.

Hegazi, A.G., El-Hady, A.F.K. (2009). Influence of honey on the suppression of human low-density lipoprotein (LDL) peroxidation (in vitro). *Evidence-based Complementary and Alternative Med* 6(1):113–121.

Irawanto, R. (2013). Distribution study of *Coryphautan lamk* from herbarium bogorienses specimens and the conservation areas in east java. In: Proceeding International Conference on Global Resource Conservation. 4(1): 135–144.

Jagannadha Rao, P.V.K., Das, M., Das, S.K. (2009). Effect of moisture content on glass transition and sticky point temperatures of sugarcane, palmyra-palm and date-palm jaggery granules. *Int J Food Sci Technol* 45(1): 94–104.

Jayanudin, J., Kurniawan, T., Kustiningsih, I. (2019). Phenolic analysis and characterization of palm sugar (*Arenga pinnata*) produced by the spray dryer. *Oriental J Chem* 35(1): 150–156.

Jnanadevan, R. (2015). Neera tapping for making coconut the most profitable horticultural crop for future. *Indian Coconut J* 22(1):20-22.

Jose, N. (2018). Neera- A potential natural health drink. *Biomedical JScientific Technical Res* 11(3):8595–8597.

Joseph, P., Gabriel, O., Richard, A., Godwin, A. (2018). Bioethanol potential from oil palm sap in Ghana. *Int J Renewable Energy Res* 4(1): 54-60.

Kamga, S.M., Brokamp, G., Cosiaux, A., Awono, A., Furniss, S., Barfod, A.S., Muafor, F.J., Gall, P.E., Sonke, B., Couvreur, T.P.(2020). Use and cultural significance of *Raphia* palms. *Econ Bot* 74: 207–225, https://doi.org/10.1007/s12231-020-09487-z.

Kapilan, R. (2015). Determination of efficient fermentation inhibitor of the tapped inflorescence sap of *Caryota urens* in Sri Lanka. *Int J Current MicrobiolApplied Sci* 4(10): 487–496.

Karamoko, D., Deni, N.D.T., Moroh, J.L.A., Bouatenin, K.M.J.P. and Dje, K.M., (2016). Biochemical and microbial properties of palm wine: Effect of tapping length and varietal differences. *Food and Nutrn Sci*7(9):763–771.

Karimi, A., Naghavi, M., Peighambari, S.A., Rasoulnia, A. (2021). Inulin content and expression of related genes in different tissues and cell suspension culture of Taraxacum kok-saghyz. *In Vitro Cellular and Dev Biol Plant* 57(6):1009-1017

Karseno, E., Yanto, T., Setyowati, R., Haryanti, P. (2018). Effect of pH and temperature on browning intensity of coconut sugar and its antioxidant activity. Food Res2(1):32–38.

Kovoor, A. (1983). The palmyrah palm potential and perspectives. FAO Plant Production and Protection Paper No 52. FAO, Rome. pp. 77.

Lasekan, O., Abbas, K.A. (2010). Flavour chemistry of palm toddy and palm juice: A review. *Trends Food SciTechnol* 21(10):494–501.

Leena, M.M., Yoha, K.S., Moses, J.A., Anandharamakrishnan, C. (2021). Electrospun nanofibrous membrane for filtration of coconut neera. *Nanotechnol Environ Eng* 6(2):24.

Lentz, D.L. (1990). *Acrocomia mexicana*: Palm of the ancient mesoamericas. *J Ethnobiol*10(2):183–194.

Lim, T.K. (2016). *Edible medicinal and non-medicinal plants* (Vol. 1, pp. 285-292). Dordrecht, The Netherlands: Springer.

Luis, G., Rubio, C., Gutiérrez, A.J., Hernández, C., González-Weller, D., Revert, C., Castilla, A., Abreu, P. and Hardisson, A., (2012). Miel de palma; composici'onnutricional de un edulcorante natural. *Nutricion Hospitalaria* 27(2): 548–552.

Makhlouf-Gafsi, I., Mokni-Ghribi, A., Bchir, B., Attia, H., Blecker, C., Besbes, S. (2016). Physico-chemical properties and amino acid profiles of sap from Tunisian date palm. *Scientia Agricola* 73(1):85–90.

Martins, A.R.O., Shackleton, C.M. (2018). The production and commercialization of palm wine from *Hyphaene coriacea* and *Phoenix reclinata* in Zitundo area, southern Mozambique. *South African J Bot*116: 6–15.

Martins, S.I.F.S., Jongen, W.M.F., Boekel, M.A.J.S. (2001). A review of Maillard reaction in food and implications to kinetic modelling. *Trends in Food SciTechnol* 11(9-10): 364–373.

Matsui, N., Okimori, Y., Takahashi, F., Matsumura, K., Bamroongrugsa, N. (2014). Nipa (*Nypafruticans*Wurmb) sap collection in southern Thailand I. Sap production and farm management. *Environ Natural Resources Res* 4(4):75-88.

May, P.H., Anderson, A.B., Balick, M.J., Fraz˜ao, J.M.F. (1985). Subsistence benefits from the babassu palm (*Orbignya martiana*). *Econ Bot* 39(2):113–129.

Meekijjaroenroj, A. (2004). Palm (Arecaceae)/pollinator interactions: Case study of two palm species, *Calamus castaneus*, *Phoenix canariensis* and floral fragrance chemistry. Montpellier University, France.

Misra, B. (2016). Neera: The coconut sap: A review. *Int J Food Sci Nutn* 1(4): 35–38.

Mogea, J., Seibert, B., Smits, W. (1991). Multipurpose palms: the sugar palm. *Agroforestry Systems* 13: 111-129.

Mummaleti, G., Sarma, C., Kalakandan, S., Sivanandham, V., Rawson, A., Anandharaj, A. (2021b). Optimization and extraction of edible microbial polysaccharide from fresh coconut inflorescence sap: An alternative substrate. *LWT* 138, 110619.

Mummaleti, G., Sarma, C., Kalakandan, S.K., Gazula, H., Sivanandham, V., Anandharaj, A. (2021a). Characterization of levan produced from coconut inflorescence sap using *Bacillus subtilis* and its application as a sweetener. *LWT* 154(2), 112697, https://doi.org/10.1016/j.lwt.2021.112697.

Naik, M., Sunil, C.K. Rawson, A., Venkatachalapathy, N. (2020). Tender coconut water: A review on recent advances in processing and preservation. *Food Reviews Int* 38:1215-1236.

Naknean, P. (2010). Factors affecting browning and crystallisation of palm sugar syrup and palm sugar cake. PhD thesis. Songkhla, Hat-Yai, Thailand: Prince of Songkla University. 336p.

Namoff, S., Veloz, A., Jim'enez, F., Rodríguez-Pe˜na, R.A., Peguero, B., Lewis. C., Moynihan, J., Abdo, M., Maunder, M., Von Wettberg. E., Meerow. A.W., Griffith. M.P., Francisco-Ortega. J. (2011). Sweet drinks are made of this: Conservation genetics of an endemic palm species from the Dominican Republic. *J Heredity*102:1–10.

Nighitha. M.T., Sudagar. I.P. (2017). Engineering properties of palmyrah palm jaggery (*Borassus flabellifer* L.). *Trends in Biosci J* 10 (43): 8974-8976.

Obahiagbon, F.I., Osagie. A.U. (2007). Sugar and macrominerals composition of sap produced by *Raphia hookeri* palms. *African J Biotechnol* 6(6): 744–750.

Orwa, C., Mutua, A., Kindt, R., Jamnadass, R., Anthony, S. (2009). Agroforestree Database: a tree reference and selection guide version 4.0 http://www.worldagroforestry.org/sites/treedbs/treedatabases.asp,

Otedoh, M.O. (1990). Sweet Raphia palm wine. *The Nigerian Field* 55:59-64.

Päivöke, A.E. (1985). Tapping practices and sap yields of the nipa palm (*Nypafruticans*) in Papua New Guinea. *Agric Ecosyst Environ13*(1): 59-72.

Patel, C., Nema, S. (2021). Sap collection, production, processing and conservation of *Caryotaurens* (Sulphi) in Bastar (Chhattisgarh). *Int J Current Microbiol Applied Sci* 10(1): 1556-1567.

Perdrix, J., Bovet, P., Larue, D., Yersin, B., Burnand, B., Paccaud, F. (1999). Patterns of alcohol consumption in the Seychelles islands (Indian ocean). *Alcohol and Alcoholism* 34(5): 773–785.

Pethiyagoda, U. (1978) Coconut inflorescence sap. Perak Planters' Association J Malaysia. pp 86-91.

Pighin, E., Díez, V.K., Di Cosimo, J.I. (2016). Synthesis of ethyl lactate from triose sugars on Sn/Al2O3 catalysts. *Applied Catalysis A: General* 517(5):151–160.

Pillai, A.R., Riyas, C.T., Sabu, K.K. (2020). A review on the unexplored and underutilized *Arenga* species in India. *Current Bot* 11:226–232.

Purnomo, H. (2007). Volatile components of coconut fresh sap, sap syrup and coconut sugar. *Int Food Res J* 14(1):45–49.

Redhead, J. (1989). Utilization of tropical foods: trees. In: FAO Food and Nutrition Paper No 47: 3. FAO, Rome, 52p.

Reinoso, K.L. (2024) Blog on Beverage Industry Data: Key Trends and Insights for 2024 and Beyond.

Rodríguez, L., Preston, T.R. (1997). Local feed resources and indigenous breeds: Fundamental issues in integrated farming systems (1). Livestock Research for Rural Dev 9(2): 32–38.

Salvi, Katewa, S.S. (2012). Chemical composition and nutritive value of sap of *Phoenix sylvestris*Roxb. *Electronic J Environ Agric Food Chem*, 578–583.

Sambou. B., Goudiaby, A., Ervik, F., Diallo, D., Camara, M.C. (2002). Palm wine harvesting by the Bassari threatens *Borassus aethiopum* populations in north-western Guinea. *Biodiversity & Conservation* 11:1149–1161.

Samuel, O., Lina, J., Ifeanyi, O. (2016). Production of vinegar from oil-palm wine using *Acetobacter aceti* isolated from rotten banana fruits. *Universal J Biomed Eng* 4(1):1–5.

Saputro, A.D., Van de Walle, D., Dewettinck, K. (2019). Palm sap sugar: A review. *Sugar Tech* 21(6):862–867.

Sheng, L., Zeng, Y., Wei, T., Zhu, M., Fang, X., Yuan, X., Luo, Y., Feng, L. (2018). Cloning and functional verification of genes related to 2-phenylethanol biosynthesis in rosa rugosa. *Genes* 9(12):576.

Shetty, P., Souza, A.D., Poojari, S., Narayana, J., Rajeeva. P. (2017). Study of fermentation kinetics of palm sap from *Cocos nucifera*. *Int J Applied Sci Biotech* 5:375-381, doi:10.3126/ijasbt.v5i3.18297.

Sola, P., Jones, D.E., Gambiza, J. (2006). Impacts of leaf harvesting and sap tapping on the ivory palm (*Hyphaene petersiana*) in south eastern Zimbabwe. *Forests, Trees and Livelihoods.* 16(4): 381-395.

Staples GW, Bevacqua RF (2006). *Areca catechu* (betel nut palm). Species profiles for Pacific Island. *Agroforest*1(13): 1-9.

Sudha, R., Niral, V., Hebbar, K.B., Samsudeen, K. (2019). Coconut inflorescence sap. *Current Science* 116(11):1809.

Sukumaran, L., Radhakrishnan, M. (2021). Effect of frozen storage on the inhibition of microbial population, chemical and sensory characteristics of coconut neera. *J Applied Microbiol* 131(4):1830–1839.

Sumadi, G. (1988). *Arenga pinnata*: a palm for agroforestry. In: Proceedings of an international workshop held in Pataya, Thailand, 2-5 November 1987, pp. 149-150.

Swamy, G.M.S. (2013). Coconut neera production and processing in Karnataka. *Indian Coconut J* 31-33.

Tapsoba, F., Savadogo, A., Zongo, C., SababenedyoTraor'e, A. (2016). Improvement of *Borassus akeassii* wines quality by controlled fermentation using *Saccharomyces cerevisiae* strains. *J Microbiol Biotechnol Food Sci* 5(6): 589–592.

Trinidad, T.P., Mallillin, A.C., Sagum, R.S., Encabo, R.R. (2010). Glycemic index of commonly consumed carbohydrate foods in the Philippines. *J Functional Foods* 2(4): 271–274.

Tuley, P. (1964) How to tap Raphia palm wine. *The Nigerian Field* 55: 59-64.

Tuley, P. (1965). Studies on the production of wine from the oil palm. *J Nigerian Institute for Oil Palm Res* 4: 284-289.

Tuley, P. (1995). The palms of Africa. Cornwall: Trendrine Press, Zennor, St Ives, Cornwall, 189p.

Van Nguyen, D., Harifara, R., Saka, S. (2016). Sap from various palms as a renewable energy source for bioethanol production. *Chem Industry and Chem Eng Quart* 22(4):355–373.

Vengaiah, P.C., Murthy, G.N.M.S., Maheswarappa, H.P. (2017). Value added food products from palmyrah palm (*Borassus flabellifer* L). *J Nutrn and Health Sci* 4(1): 2–5.

Vengaiah, P.C., Ravindrababu, D., Murthy, G.N., Prasad, K.R. (2013). Jaggery from palmyrah palm (*Borassus flabellifer* L.)- Present status and scope. *Indian J Traditional Knowledge* 12(4):714–717.

Vichi S. (2010). Extraction techniques for the analysis of virgin olive oil aroma. Preedy, V.R., Watson, R.R. (eds.), Olives and Olive Oil in Health and Disease Prevention, Academic Press, San Diego, pp. 615-623.

Victor, I.R.M. (2015). Processing of *Arenga pinnata (palm)* sugar. PhD Thesis. Qu'ebec, Canada: Department of Bioresource Engineering Faculty of Agricultural and Environmental Sciences Macdonald Campus of McGill University Sainte-Anne-De-Bellevue. pp 194.

Wang, Y., Zhao, D., Rodríguez-Padr'on, D., Len, C. (2019). Recent advances in catalytic hydrogenation of furfural. *Catalysts* 9(10): 796, https://doi.org/10.3390/catal9100796.

Widyaningrum, T., Suharjono, Ardyati, T., Aulanni'Am (2019). Screening and identification indigenous yeast from neera Siwalan for bioethanol production. *IOP Conference Series: Earth and Environmental Science* 230(1),012073.

Xia, Q. (2011). Chemical composition changes of post-harvest coconut inflorescence sap during natural fermentation. *African J Biotechnol* 10(66):14999–15005.

Yusoff, N.A., Yam, M.F., Beh, H.K., Razak, K.N.A., Widyawati, T., Mahmud, R., Ahmad, M. and Asmawi, M.Z. (2015). Antidiabetic and antioxidant activities of *Nypa fruticans* Wurmb. vinegar sample from Malaysia. *Asian Pacific J Tropical Med* 8(8): 595–605.

Chapter 2

Palm Sap Yield: The Effects of Coconut Genotypes and Seasons

R. Sudha,
and V. Niral*
ICAR-Central Plantation Crops Research Institute, Kasaragod, Kerala, India

Abstract

Coconut inflorescence sap, also known as *neera*, is a sweet, translucent juice extracted from the unopened inflorescence of coconut trees. It is oyster white in color and offers high nutritive value. Research has demonstrated that both sap yield and quality parameters vary with coconut genotypes and seasonal conditions. Thus, understanding the effects of genotypes and season on coconut inflorescence sap yield and quality is crucial. Among the genotypes, Jamaica Tall yielded the highest amount of inflorescence sap (3.60 l day^{-1}), followed by Orissa Tall (3.05 l day^{-1}). Regarding seasonal variations, the highest sap yields were recorded during the post-monsoon (October-November) and monsoon (June-September) periods, while the lowest yields occurred during the winter season (December-February) across all the genotypes. The quality of inflorescence sap varied significantly among different coconut accessions in terms of total soluble solids (TSS), total sugars, reducing sugars, phenolics, amino acids, ascorbic acid, and protein contents. The winter season recorded the appreciable and desired quality parameters of sap across all genotypes. Overall, the chapter highlights the significant influence of genotypes and seasons on both the yield and quality of coconut inflorescence sap.

* Corresponding Author's Email: niral.v@icar.gov.in

In: Coconut Palm Sap
Editors: K. B. Hebbar, V. Krishnakumar and S. V. Ramesh
ISBN: 979-8-89530-640-6

Keywords: seasonal influence, coconut genotypes, quality parameters, sap yield, mineral nutrition

1. Introduction

Coconut inflorescence sap is a natural, unfermented, and healthy beverage obtained by tapping the unopened immature inflorescence of the coconut palm (Borse et al., 2007). The sap is a sweet, translucent juice, oyster white in color, with high nutritional value. It is a rich source of reducing and non-reducing sugars as well as essential minerals and vitamins, including iron, phosphorus, and ascorbic acid. One of its most significant characteristics is the low glycemic index (GI), which makes it suitable for diabetic patients (Manohar et al., 2007). Recently, there has been a surge in global demand for low GI sugars. Additionally, other products such as coconut sugar, jaggery, and honey derived from the unfermented sap also offer potential benefits to the coconut farmers.

The yield of coconut inflorescence sap is highly variable and depends on factors such as palm variety, vigor, and climatic conditions. The nutritional status of the palms also significantly impacts sap yield. Studies have shown that different coconut genotypes vary in sap yield and quality (Joseph et al.2018; Samsudeen et al., 2013; Ysidor et al., 2014). The sap production *in vivo* involves different processes such as hydro-mineral absorption, photosynthesis, and transpiration, and is transported through the plant's vascular tissues - phloem and xylem - according to the plant's needs (Köstner et al., 2008; Nakamura et al., 2004; Ndon 2003). Thus, the sap yield per coconut palm is directly related to its genetic potential. Ongoing research and breeding programs are focused on developing and selecting cultivars with high sap yields.

2. Inflorescence Sap Yield

The yield of sap from coconut palms varies significantly among different cultivars, influenced by genetic characteristics and growth conditions. Some cultivars are naturally high sap yielders due to their genetic makeup. Additionally, the age and maturity of the palm can affect sap yield, as cultivars reach optimal sap production at different stages. The flowering frequency of cultivars also impacts sap yield; some may produce more inflorescences or have a higher sap flow rate per inflorescence. Other factors, though not directly related to the cultivar, interact with it to influence sap yield. These include soil type, water availability, and climate. Some cultivars may be better adapted to specific environmental conditions, leading to higher sap yields under favorable conditions.

Secretaria et al., (2002) reported significant differences in sap and nut yields among coconut varieties. A study in the Philippines on the Sequential Coconut Toddy and Nut Production (SCTNP) scheme, found that dwarf x tall hybrids viz., MRD x BAYT (PCA 15-3), MRD x TAGT and CATD x LAGT and a local tall cultivar, Laguna (LAGT), produced significantly higher sap and nut yields compared to the Catigan Dwarf hybrid (CATD x CATD), and Tacunan Dwarf cultivar (TACD). Further, among the hybrids and tall variety, MRD x BAYT produced significantly high sap yield (90 l $palm^{-1}$ $month^{-1}$) followed by CATD x LAGT, MRD x TAGT and the least volume of sap yield was recorded in LAGT. Rethinam et al., (2002) also observed variations in sap yield among different genotypes, including Laccadive Ordinary Tall, Andaman Ordinary Tall, Laccadive Micro Tall, and West Coast Tall.

Legaspi (2006) studied sap yield in tall and dwarf coconut varieties, including Salambuyan, San Ramon, Malayan Red Dwarf, and Santo Nino Green Dwarf. The tall palms exhibited superior total toddy yields compared to dwarfs. Variability in toddy yield per spadix was significant, with factors such as tapping duration, spadix circumference, and length of spadix correlating positively with yield, while the number of leaves and vascular bundles correlating negatively. According to Wasantha et al., (2009), sap flow in coconut palms depends on the variety and soil conditions. The sap yield can range from 2 to 4.5 litres depending on palm health and management practices (Muralidharan and Deepthi 2013) (Figure 1).

Konan et al., (2013) evaluated various coconut hybrids and varieties and found that the hybrid PB 113+ produced the highest sap volumes per spadix, followed by PB 121+. The West African Tall (WAT) and Malayan Yellow Dwarf (MYD) varieties produced the lowest sap volumes. Samsudeen et al., (2013) observed that the highest average sap production was 960 ml day^{-1} in West Coast Tall (WCT), followed by 880 ml in the CODxWCT hybrid. The lowest production was 104.2 mL in Chowghat Orange Dwarf (COD). The average sap production was statistically comparable among WCT, LCT, and CODxWCT hybrid, all significantly higher than COD.

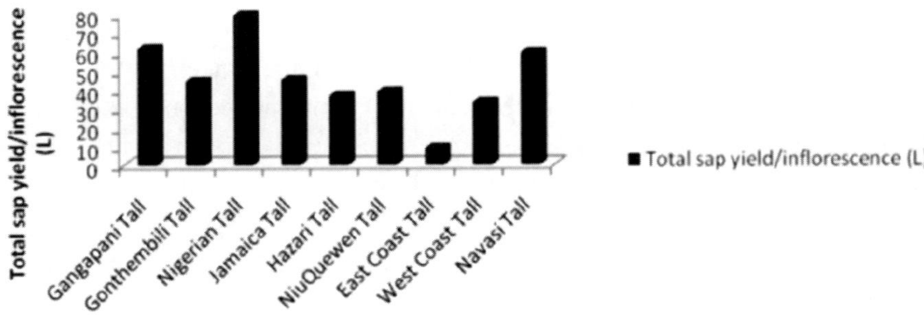

Source: Sudha et al., 2021.

Figure 1. Variability in sap yield per inflorescence of different coconut genotypes.

At ICAR - CPCRI, Kasaragod, India 14 accessions when evaluated for inflorescence sap production, indicated variation in sap quality and quantity. The highest sap yield was recorded in Laccadive Ordinary Tall, followed by Nigerian Green Dwarf (though considered as a dwarf accession, it has relatively a taller plant habit), Federated Malay States Tall, WCT and Fiji Tall. In general, sap yield was observed to be less in dwarfs. Among the dwarf accessions, sap yield was low in COD followed by Gangabondam, whereas, MYD and Cameron Red Dwarf produced relatively high quantity of sap (Anon. 2016). Tulalo and Mawardi (2018) studied sap yield in dwarf varieties and found that Waingapu Red Dwarf (WRD) produced more sap (1,007 ml) compared to Salak Green Dwarf (SGD) (741 ml) and Raja Brown Dwarf (RBD) (628 ml^{-1} palm day^{-1}).

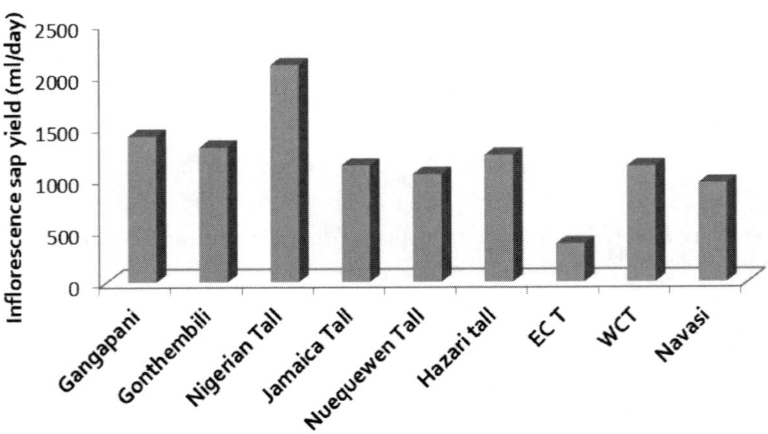

Figure 2. Inflorescence sap yield in different coconut genotypes (Sudha et al., 2019a).

Sudha et al., (2019a) found that the Nigerian Tall recorded the highest total sap yield, significantly outperforming other genotypes viz., Jamaica Tall, Hazari Tall, Navasi Tall, Gonthembili Tall, Gangapani Tall, West Coast Tall, East Coast Tall and Niu Quewen Tall. East Coast Tall recorded the lowest sap yield. The highest average sap yield (ml day^{-1}) was recorded by Nigerian Tall (2098.74) followed by Gangapani Tall (1410.47) and Gonthembili Tall (1303.33) (Figure 2).

Raghu and Joseph (2020) reported that the variety WCT produced the highest inflorescence sap yield of 3.14 l day^{-1}, followed by the hybrids: Kerasree (2.09 l day^{-1}) and Keraganga (1.80 l day^{-1}). The MYD gave the lowest yield at 0.84l day^{-1}. Novarianto et al., (2021) studied dwarf coconut varieties in North Sulawesi, Indonesia, and found that Waingapu Red Dwarf (WRD), Salak Green Dwarf (SGD), and Nias Yellow Dwarf (NYD) had the highest sap and sugar yields. Sap production ranged from 1.1 to 3.3 l $palm^{-1}$ day^{-1}, with potential sugar yields of 2.09 tons, 1.64 tons, and 1.56 tons $month^{-1}$ ha^{-1}, respectively. Legaspi (2024) estimated sap yield from four local dwarf varieties in the Philippines. Aromatic Green Dwarf (AGD), Equatorial Green Dwarf (EGD), San Isidro Green Dwarf (SGD), and Talisay (TAL). AGD produced the greatest volume of sap followed by EGD, SGD while TAL produced the least.

3. Duration of Sap Production

In healthy coconut palms, tapping can be continued almost without any break throughout the year. If the trees are well-maintained and cared for, and the tappers follow proper tapping techniques, the palms can produce sap consistently. The duration of sap production, or the period during which a coconut palm continues to yield sap, varies among different cultivars. This variation is influenced by genetic factors, management practices, and environmental conditions. Generally, hybrids or improved varieties offer extended sap production periods and higher yields compared to traditional or less specialized cultivars.

According to Nathanael (1955), the daily average yield $palm^{-1}$ continues to increase steadily up to the 10th month. This suggests that palms become more productive with consistent tapping, likely due to physiological adjustments in the trees as they adapt to the extraction process. After the 10th month, there is only a slight decrease in yield. This means that tapping could, in theory, continue for a full 12 months without significant loss in productivity, but this slight drop indicates that the trees may be nearing the point of needing recovery. It was also highlighted that the process of tapping can be followed in rotation and continuous tapping as detailed below.

3.1. Tapping Different Sets of Palms in Rotation

If a rotation system is used, where different groups of palms are tapped annually while others rest; it is both safe and economical to tap each set of palms for a full 12-month period. This system ensures that some palms are always at rest, allowing for recovery and sustaining productivity over time. This rotational method is ideal for plantation-style operations, where large number of palms are available, enabling the division of trees into different sets for tapping at different times.

3.2. Continuous Tapping on the Same Palms

If the same palms are tapped continuously without rotation, a 12-month tapping cycle might not be sustainable in the long run due to the stress on the trees. In such cases, the present system of eight months of tapping followed by a four-month rest period is recommended. This system gives the palms sufficient time to recover, ensuring they remain healthy and productive for many years.

Generally tapping is practiced for six months in a set of trees. If the tree is vigorous, tapping may be prolonged for even two years. In countries like the Philippines, Sri Lanka and Pacific Islands, palms are continuously tapped for a good number of years. Some trees have been *neera* yielders throughout their life period. In Sri Lanka, coconut palms could be tapped for indefinite number of seasons as long as the rainfall was sufficient and the palms were rested for four months in a year. In the drier areas however, the spathes after 3-5 consecutive seasons of tapping were reported to progressively diminish in size and eventually abort. In Thailand, tapping was generally sustained for 10-11 months (Menon and Pandalai 1958). Naik et al., (2013) suggested that a palm can be tapped for six months at a stretch in a year and allowed for the next six months for tender nut or coconut production as a health restoration measure for the palm.

The duration of sap production depends on the length of the spadix and the consistency of sap flow during the tapping cycle. Cortazar et al., (2010) reported that the duration of inflorescence sap production is about 20-30 days spadix^{-1}. Konan et al., (2013) assessed sap production from spadices of four coconut cultivars in Côte d'Ivoire and found that the hybrid PB 113+ had the longest sap production duration of 46.78 days, while MYD had the shortest duration of 24 days. Samsudeen et al., (2013) observed that the sap production duration for West Coast Tall (WCT), Laccadive Ordinary Tall, Chowghat Orange Dwarf, and COD × WCT ranged from 60 to 70 days. Raghu and Joseph (2020) investigated the influence of varieties viz., Malayan Yellow Dwarf, Kerasree, West Coast Tall, and Keraganga on sap yield and reported tapping duration ranging from 40 to 60 days.

The number of days taken for the initiation of sap flow differed with genotypes, which varied from 16.1 to 23.5 days. Among the genotypes, Nigerian Tall started the flow sap from 16.1 days onwards whereas East Coast Tall took 23.5 days to start the flow (Sudha et al., 2019a). The duration of sap production differed significantly in all the genotypes evaluated which varied from 34 to 56 days. Spathes of Nigerian Tall produced sap for the maximum duration of 56 days than other genotypes, while the least duration of sap production was recorded in East Coast Tall (34 days). Continuous tapping of coconut palms can be practiced for up to six months. However, after this period, a reduction in sap yield was observed in various genotypes.

Novarianto et al., (2021) found that the duration of tapping of dwarf coconut varieties in North Sulawesi, Indonesia, ranged from 31.42 to 43.21 days spadix^{-1}, with Waingapu Red Dwarf (WRD) having the longest tapping time of 43.21 days and Salak Green Dwarf (SGD) the shortest (31.42 days). Legaspi (2024) reported that among the four dwarf varieties evaluated, Aromatic Green Dwarf showed the longest tapping duration during the dry season (40.34 days), comparable to Salak Green Dwarf (38.21) and Equatorial Green Dwarf (31.55) during the dry season. During the wet season all the three varieties showed a tapping duration of 28 days. The shortest tapping duration recorded was for Talisay in both the seasons (12.78 days in dry and 20.88 days in wet season).

4. Biochemical Features of Coconut Sap : Effect of Genotypes

The biochemical characteristics of coconut inflorescence sap can vary significantly among the cultivars, affecting its sweetness, nutritional value, and suitability for various uses such as fresh consumption or production of value-added products. Understanding these characteristics is crucial for selecting the most suitable cultivars for specific purposes. The biochemical composition, particularly sugar content and acidity, influences the shelf life and stability of the sap.

The chemical composition of coconut inflorescence sap reflects the composition of phloem sap from the source organs (MacRobbie 1971). Nakamura et al., (2004) noted that the exudates collected from the cut end of a flower-bearing stem, derived from sieve tubes, is likely to be similar to phloem sap. Phloem sap is a complex mixture of organic and inorganic substances, with sugars and amino acids being principal metabolites (Komor et al., 1996). The sap is typically moderately alkaline, with previous reports of acidic pH likely due to contamination with xylem exudates and bark tissue (Dinant et al., 2010).

Legaspi (2006) analyzed biochemical parameters of sap derived from two tall and two dwarf coconut varieties. The chemical compositions of the sap from both palm types were comparable, except for the phosphorus and total nitrogen contents, which were higher in tall palms and dwarf palms, respectively. The total soluble solids (TSS) content was higher in the tall palm Salambuyan compared to dwarf palms during the wet season. Samsudeen et al., (2013) analyzed biochemical parameters of three varieties (WCT, LCT, COD) and one hybrid (CODxWCT). Among the varieties, the highest total sugar content was recorded in West Coast Tall (WCT), followed by Laccadive Tall (LCT). There were no significant differences in reducing sugar and protein content among the varieties tested, though reducing sugar was slightly higher in CODxWCT and LCT.

Konan et al., (2013) conducted chromatographic analysis of inflorescence sap from four coconut varieties. Sucrose content (g 100 ml^{-1}) varied from 9.40 to 12.24, with the highest content in MYD (12.24) and the lowest in WAT (9.40). The hybrids PB 121+ (10.31 g 100 ml^{-1}) and PB 113+ (10.38 g 100 ml^{-1}) had intermediate sucrose levels. Glucose content ranged from 1.63 to 1.84 g 100 ml^{-1}, while fructose ranged from 1.24 to 1.52 g 100 ml^{-1}, with no significant differences among the cultivars. MYD had the highest glycerol content (0.094 g 100 ml^{-1}), while the other cultivars had similar glycerol levels (0.05 to 0.058 g 100 ml^{-1}). MYD also showed higher mannose content (0.019 g 100 ml^{-1}) compared to WAT (0.011 g 100 ml^{-1}), PB 121+ (0.008 g 100 ml^{-1}), and PB 113+ (0.014 g 100 ml^{-1}). The sweetness index was the highest in dwarf coconuts compared to hybrids PB 121+ (13.68), PB 113+ (13.56), and Tall WAT (13.03).

Sudha et al., (2021) found significant differences in TSS and total sugar content among nine genotypes of coconut. The highest TSS content was recorded in WCT, while the lowest was in East Coast Tall. West Coast Tall also had the highest total sugar content (13.4 g 100 ml^{-1}), followed by Gangapani Tall (12.75 g 100 ml^{-1}), with the lowest in Nigerian Tall (12.15 g 100 ml^{-1}). West Coast Tall also showed the highest protein and amino acid contents.

Raghu and Joseph (2020) reported that WCT exhibited the highest reducing sugar content (0.52 g 100 ml^{-1}), comparable to Kerasree (0.51 g 100 ml^{-1}). Keraganga (0.43 g 100 ml^{-1}) and MYD (0.40 g 100 ml^{-1}) showed similar reducing sugar contents. MYD had the highest non-reducing sugar content (10.23 g 100 ml^{-1}), comparable to WCT, while Kerasree had the lowest non-reducing sugar content (9.91 g 100 ml^{-1}), similar to Keraganga. The highest total sugar content (10.66 g 100 ml^{-1}) was in WCT, similar to MYD and Keraganga, while Kerasree had the lowest content (10.42 g 100 ml^{-1}). Total poly phenol content ranged from 4.02 mg 100 ml^{-1}(MYD) to 5.71 mg 100 ml^{-1}(WCT), with WCT having the highest value, comparable to MYD and Kerasree. These varieties did not differ significantly in terms of pH and phenolic content of the sap. However, WCT and MYD were superior in vitamin C content.

Okoma et al., (2020) characterized crystalline sugars from inflorescence sap of coconut varieties viz., PB 113+, PB 121+ and GOA (Grand Ouest Africain) in Ivory Coast. Chromatographic analysis, reported similar contents of sucrose, glucose, and fructose among the saps of these varieties. However, the sweetness index of sugars was the highest in sap from PB 121+ (90.10% ± 2.4%), followed by GOA (89.60% ± 1.79%) and PB 113+ (89.18% ± 2.02%). Quality of sap differed among varieties, with the sap of the hybrid PB 113+ being vitamin C-rich; PB 121+ having higher proportions of vitamins B1 and B6; GOA has predominantly higher vitamin B2. The sap of the high yielding hybrid PB 113+ was found to contain 43mg 100 $ml^{-1}$of vitamin C, 27 mg 100 $ml^{-1}$of vitamin B1, 2 mg 100 $ml^{-1}$of vitamin B2, and 19.1 mg 100 $ml^{-1}$of vitamin B6.

5. Mineral Nutrients of Coconut Sap : Genotypic Differences

Different coconut genotypes have distinct genetic backgrounds that influence their mineral uptake and accumulation. Understanding these differences is crucial for selecting coconut genotypes based on desired sap characteristics, whether for fresh consumption, health benefits, or industrial applications.

Joelle et al., (2017) analysed 13 mineral nutrients in coconut sap, including eight macroelements (K, Cl, Si, Na, Mg, P, S, and Ca) and five trace elements (Fe, Cu, Mn, Zn, and Br). Sap from the genotype MYD showed the highest mineral content, with significant levels of K, P, Fe, Si, Na, S, and Mg. The PB113+ hybrid also exhibited high levels of minerals. In contrast, the sap from WAT had lower mineral contents. Raghu and Joseph (2020) investigated the nutritional qualities and total mineral content in the inflorescence sap of four coconut varieties: WCT, MYD, Kerasree, and Keraganga. They found no significant differences in protein nitrogen and non-protein nitrogen content among the varieties. Phosphorus content ranged from 0.43% in MYD to 0.26% in Kerasree, while potassium content was the highest in WCT (1.40%) and the lowest in MYD (1.18%). MYD had the highest calcium content at 579.20 mg l^{-1}. Magnesium content did not vary significantly among the varieties. The highest sulfur content was observed in WCT (0.30%), and the highest copper content was found in MYD (2.31 mg l^{-1}). Zinc, iron, and manganese content did not show significant variations among the different varieties. Overall, the total mineral content of sap was the highest in WCT, followed by MYD.

6. Morphological Traits Influencing Sap Production

Several factors influence the sap yield of coconut palms. Among these factors, the vegetative and floral characteristics of the palms - such as leaf production, the number of living fronds, and the structure of the spadix (circumference and length) - significantly contribute to variations in sap yield among the coconut hybrids and cultivars (Secretaria et al., 2002). Palms with a greater number of living fronds tend to yield more sap, as each living frond can support a spadix suitable for tapping. Additionally, the floral structure, particularly the length of the inflorescence, is a good indicator of higher sap yield. Longer inflorescences generally contain more spikelets and flowers, which can increase sap production.

In Sri Lanka, traditional tappers use criteria such as a higher number of leaves and long, relatively thin inflorescence sheaths and identify palms for tapping (Ranasinghe2007). These characteristics facilitate the bending of flower stalks without splitting, ensuring sustainably high sap yields. Konan et al., (2013) found that the duration of sap production was positively correlated with length of the spadix and their median circumference. Also, the sap yield per inflorescence was linked to duration of sap production, and spadices that tended to produce more fruits also provided higher sap yields.

Legaspi (2006) investigated the effect of varieties (Salambuyan Tall, San Ramon Tall, Malayan Red Dwarf, and Santo Nino Green Dwarf) on the morpho-anatomy of the spadix used for tapping. Most developing female flowers and cut spikelets were located in the middle section of the spadix. The tip section had smaller circumference, fewer, shorter, and narrower vascular bundles. Dwarf palms showed a shorter interval from the first pairing to the commencement of toddy flow compared to tall palms. The bulk of the sap yield (80% of the total) came from the tip and middle sections of the spadix. Sap flow increased continuously within the tip section, peaked in the middle section, and stopped at the base. Sap yield was positively correlated with tapping duration, circumference of the spadix per section, length of the tapped spadix per section, and cut spikelets per section. Conversely, sap yield was negatively correlated with the number of leaves and the total number, width, and length of the vascular bundles. On the other hand, Secretaria et al., (2002), based on correlation among coconut varieties, particularly in CATD x LAGT hybrid, observed that palms with more leaves on the crown showed tendency to yield higher sap. Every living frond of bearing palms has the capacity to support a spadix which can be used for tapping. The relationship between the morphological traits and sap yield was assessed by correlation analysis (Table 1). The sap yield showed positive correlation with parameters such as number of leaves, stem girth, length of the inflorescence and median circumference of the inflorescence while it was negatively correlated with petiole length and intermodal length (Sudha et al., 2021).

Table 1. Correlation analysis of average yield of inflorescence sap with sap production parameters (Sudha et al., 2021)

	No. of leaves	Petiole length	Internodal length	Stem girth	Length of inflorescence	Median circumference of inflorescence	Sap yield
No. of leaves	1						
Petiole length	-0.422	1					
Internodal length	-0.635	-0.144	1				
Stem girth	0.681*	-0.432	-0.661	1			
Length of inflorescence	-0.159	0.470	0.477	-0.225	1		
Median circumference of inflorescence	0.553	0.014	-0.243	0.068	-0.038	1	
Sap yield	0.762*	-0.356	-0.510	0.398	0.106	0.438	

* Significant at 5% level.

Legaspi (2024) reported that dwarf coconut varieties exhibited different lengths of tappable spadix portions regardless of the tapping season. Mean comparisons revealed that the Talisay variety had the longest tappable portion, comparable to Equatorial Green Dwarf. The shortest tappable portion was recorded for Aromatic Green Dwarf, which was comparable to Salak Green Dwarf. The SGD variety had a tappable portion comparable to EGD, while ARO had the shortest discarded tip of the tappable spadix, and SGD had the longest discarded tip.

7. Seasonal Influence of Sap Production in Different Genotypes

Tapping of coconut palms can be conducted year-round; however, the yield of sap varies with the seasons. In India, there are two major tapping periods: the dry season (November-March) and the wet season (April-October). Typically, tapping is practiced for six months on a set of trees, but if the tree is vigorous, tapping may extend for up to two years. In countries such as the Philippines, Sri Lanka, and the Pacific Islands, palms are often tapped continuously for many years, and some trees can produce sap throughout their entire lifespan.

Legaspi (2006) examined the seasonal effects on sap production across four coconut varieties. The study found that spadices tended to be longer during the dry season. Although more leaf production occurred during the wet season, the sap yield was negatively correlated with number of leaves. The circumference of the spadix was larger at the middle and base sections during the dry and wet seasons, respectively. Tapping duration was longer in the dry season, and the TSS content was higher in Salambuyan Tall compared to the dwarf palms during the wet season. The total sugar content of the sap was generally higher in the wet season.

Samsudeen et al., (2013) evaluated sap yield from three coconut varieties namely, West Coast Tall (WCT), Laccadive Ordinary Tall (LCT), and Chowghat Orange Dwarf (COD) and one hybrid, CODxWCT (Chandra Sankara), across four seasons: monsoon, post-monsoon, winter, and summer. The optimal time for tapping is from the middle of the monsoon to the beginning of winter. It is also recommended to avoid tapping during the summer months under rainfed conditions. In another study on influence of seasons on sap yield of different coconut genotypes, Sudha et al., (2019b) reported the highest sap yield during the post-monsoon (October-November) and monsoon (June-September) seasons, while the lowest yield occurred in the winter season (December-February). However, the sap collected in winter showed the good quality parameters across all varieties. Legaspi (2024) assessed seasonal effects on sap yield in four dwarf varieties (Aromatic Green Dwarf (AGD), Equatorial Green Dwarf (EGD), San Isidro Green Dwarf (SGD), and Talisay (TAL)) in the Philippines. The study observed that AGD produced the highest volume of sap during the wet season. Generally, more sap is produced in the wet season compared to the dry season. Seasonal variations significantly affect the length of the spadix, age of the spadix at tapping, and the timing of various tapping stages. There was also an interaction effect between season and variety on spadix age at tapping, tapping duration, and sap volume produced.

8. Storage of Coconut Sap : Varietal Influence

Coconut inflorescence sap is highly perishable, making it essential to understand the optimal storage conditions and the influence of variety on its properties for both commercial and local consumption purposes. Sap from some coconut varieties may exhibit better storage characteristics due to their inherent biochemical composition or physical properties.

Konan (2015) investigated the changes in physicochemical parameters of inflorescence sap from four varieties (Malayan Yellow Dwarf (MYD), West African Tall (WAT), PB 121+, and PB 113+) during storage. The study revealed significant decreases in total sugar content (from 16.08 to 0.93 g100 ml^{-1}), pH (from 7.32 to 4.17), and relative density (from 1.055 to 1) across all varieties. Additionally, during storage, TSS and dry matter content dropped from 16.31to 3.82% and 16.47 to 3.09 g $100g^{-1}$, respectively. Conversely, total titratable acidity and tannin content increased significantly, with total titrable acidity rising from 0.26 to 1.85 g 100 g^{-1} and tannins from 5.65 to 13.2 mg l^{-1}. Reducing sugars and protein contents showed partial increases, from 6.60 to 8.29 g 100 $ml^{-1}$and 2.34 to 2.84 g 100 ml^{-1}, respectively, after 24 h of storage. Total polyphenol content peaked at 606.84 to 1316.24 mg l^{-1}after 36 h, before declining. Notably, WAT differed from the other varieties with the highest content of total polyphenols and tannins during the 120 h of storage.

In India, Singaravadivel et al., (2012) analyzed the physicochemical changes like pH, turbidity, acidity, alcohol content, total sugars, and TSS, reducing sugar and non-reducing sugar in coconut sap on storage in local coconut palms. They reported change in pH from 5.5 to 4.21 and increase in the titratable acidity from 0.75 to 0.91 and also alcohol content from 0.2 to 4.5% on storage for 24 h at 28 ± 2°C. The ash and ascorbic acid content of

0.12% and 3.5 mg 100ml^{-1} in fresh sap increased upon storage to 0.38% and 5.8 mg 100 ml^{-1} respectively. The total carbohydrate content of 66 mg 100 ml^{-1} reduced to 40 mg 100 ml^{-1} after 24 h of storage. Raghu and Joseph (2020) examined biochemical changes in sap stored for one week from varieties including MYD, Kerasree, WCT, and Keraganga. The biochemical properties - pH, electrolyte concentration, reducing sugars, non-reducing sugars, total sugars, alcohol, phenol, and vitamin C, of sap stored under refrigerated conditions, were assessed on the 1st, 3rd, and 7th day of storage. The study found significant reductions in pH, total sugars, reducing sugars, and non-reducing sugars, while increases in alcohol, vitamin C, and phenol content were observed across all varieties. Similarly in sap of the WCT coconut, due to the process of fermentation, pH of the sap and total sugar content declined rapidly on storage under ambient conditions, while acidity, turbidity, and reducing sugar content significantly increased. Similarly, the vitamin C and total sugar degradation also increased with the atmospheric fermentation (Pandiselvam et al., 2021).

Conclusion and Perspectives

The genotypes of coconut palm significantly impact the characteristics of inflorescence sap, including its yield, biochemical composition, flavour, aroma, and storage stability. Genotypes with high sap yield, superior biochemical attributes, and enhanced storage stability are particularly valued in commercial sap production. By understanding these genotype-specific differences, growers can select varieties/cultivars that meet specific needs—whether for commercial production or local consumption. This knowledge supports ongoing efforts to enhance sap quality and sustainability in coconut farming, ultimately benefiting both producers and consumers.

References

Anon (2016) *Annual Report of ICAR-CPCRI*, Kasaragod, Kerala, India, 2016.

Borse, B.B., Rao, L.J.M., Ramalakshmi, K., Raghavan, B. (2007). Chemical composition of volatiles from coconut sap (neera) and effect of processing. *Food Chem* 101: 877-880. http://dx.doi.org/10.1016/j.foodchem.2006.02.026.

Cortázar, R.M., Rogelio, F.F., Fuentes del, A.I.M. (2010). Procesoproductivo de la ''tuba'' de coco – unanuevaalternativaeconomica para loscococultores del surestemexicano, Centro de investigacion regional sureste. *Campo Experimental Chetumal* 1, 43.

Dinant, S., Bonnemain, J.L., Girousse, C., Kehr, J.(2010). Phloem sap intricacy and interplay with aphid feeding, *Comptes Rendus Biologies* 333(6–7): 504-515, https://doi.org/10.1016/j.crvi.2010.03.008.

Joelle, O.D.M., Ysidor, K.G., Louis, K.K.J., Rachel, A.R., Marius, B.G.H. (2017) Mineral nutrients in inflorescence sap produced from various coconuts (*Cocos nucifera* L.) cultivated in Côte d'Ivoire. *J Exp Agric Int* 16(2): 1-9, doi: 10.9734/JEAI/2017/32557.

Joseph, B., Reghu, R.S., Gladis, R., Aparna, B. (2018). Influence of coconut varieties on yield and quality of coconut inflorescence sap-neera. *Indian J Sci Res* 19(1): 57–60.

Komor, E., Orlich, G., Weig, A., Köckenberger, W. (1996). Phloem loading-not metaphysical, only complex: towards a unified model of phloem loading. *J Exp Bot* 47:155–1164, https://doi.org/10.1093/jxb/47.Special_Issue.1155.

Konan, N.Y. (2015). Valorization of coconut sap (*Cocos nucifera L.*) in Côte d'Ivoire: Evaluation of the production and physico-chemical characterization of the sap of inflorescences of four coconut cultivars. Doctoral thesis from the University Félix Houphouet Boigny, Abidjan, Côte d'Ivoire, 170 p.

Konan, N.Y., Konan, K.J.L., Assa, R.R., Konan, B.R., Okoma, D.M.J., Allou, K., Biego, G.M.H. (2013). Assessment of sap production parameters from spathes of four coconut (*Cocos nucifera* L.) cultivars in Côte d'Ivoire. *Sustain Agric Res* 2(4):87–94.

Köstner, B., Matyssek, R., Heilmeier, H., Clausnitzer, F., Nunn, A.J., Wieser, G. (2008). Sap flow measurements as a basis for assessing trace-gas exchange of trees. *Flora* 203:14-33, http://dx.doi.org/10.1016/j.flora.2007.09.001.

Legaspi, S. (2006). Yield and quality of toddy from tall and dwarf varieties of coconut (*Cocos nucifera L.*) as influenced by inflorescence morpho-anatomy and season. University of the Philippines Los Baños (UPLB). Doctoral Thesis Dissertation.

Legaspi, S. (2024). Floral attributes and toddy yield of selected dwarf coconut Varieties as affected by tapping season. *Sustainable Energy and Environment Review* 2(1):1-7.

MacRobbie, E.A.C. (1971). Phloem translocation. Facts and mechanisms: A comparative survey. *Bioi Rev* 46:429-481.

Manohar, E.C., Kindipan, N.M.L., Sancha, L.V. (2007). Coconut sap sugar production: From farm to market and from wealth to health. *Philipp. Entomologists* 32(1):20.

Menon, K.P.V., Pandalai, K.M. (1958). *The Coconut Palm – A Monograph*, Times of India Press, Bombay, India.

Muralidharan, K., Deepthi, N.S. (2013). Coconut neera- the hidden unexplored treasure. *Indian Coconut J* p 4-8.

Naik, J.B., Suresh, P.R., Manjusha, M., Balachandran, P .V., Balakrishnan, P.C. (2013) 'Keraamritham'-a health drink from coconut inflorescence sap. *Indian Coconut J* 56(1):9–10.

Nakamura, S.I., Watanabe, A., Chongpraditnum, P., Suzui, N., Hayashi, H. (2004). Analysis of phloem exudate collected from fruit-bearing stems of coconut palm: Palm trees as a source of molecules circulating in sieve tubes. *Soil Sci Plant Nutn* 50(5):739-745, http://dx.doi.org/10.1080/00380768.2004.10408530.

Nathanael, W.R.N. (1955). Toddy yield from coconut palms in Ceylon. *Ceylon Coconut Quart* 6: 8–16.

Ndon, B.A. (2003). The Raffia palm. (1st ed.), Concept Publications Ltd, Lagos, Nigeria. p. 16.

Novarianto, H., Tulalo, M.A., Mawardi, S. (2021). IOP Conf. Series: Earth and Environmental Science 800, 012024, doi:10.1088/1755-1315/800/1/012024.

Okoma, D., Muriel, J., Konan, K., Louis, J., Rebecca, A.R. (2020). Sensory properties of table sugars derived from the inflorescences sap of three coconut (*Cocos nucifera* L.) cultivars in Côte d'Ivoire. *Am J Food Nutrn* 8(3):90-100, doi: 10.12691/ajfn-8-3-6.

Pandiselvam, R., Manikantan, M.R., Binu, S.M. (2021). Reaction kinetics of physico-chemical attributes in coconut inflorescence sap during fermentation. *J Food Sci Technol* 58, 3589–3597, https://doi.org/10.1007/s13197-021-05088-3.

Raghu, R.S., Joseph, B. (2020). Variability among the coconut varieties with respect to nutritional qualities and total mineral content in inflorescence sap. *Int J Chem Stud* 8(2):2231-2234, doi: 10.22271/chemi.2020.v8.i2ah.9080.

Ranasinghe, C.S., Silva, L.R.S. (2007). Photosynthetic assimilation, carbohydrates in vegetative organs and carbon removal in nut producing and sap-producing coconut palms. *Cocos* 18:45-57.

Rethinam, P., Sreekumar, P., Nanda Kumar, T.B. (2002) Coconut Products, Niseema Printers, Kochi.

Samsudeen, K., Niral, V., Jerard, B.A., Berwal, M.K., Sugatha, P.P., Hebbar, K.B. (2013). Influence of variety and season in the inflorescence sap production in *Cocos nucifera* L. *J Plant Crops* 47(1): 57–61.

Secretaria, M.I., Ebuna, R.M., Magat, S.S., Maravilla, J.N., Santos, G.A., Baylon, G.B. (2002). Performance of selected coconut varieties/hybrids under sequential coconut toddy and nut production scheme or SCTNP. *CORD* 18: 34. 10.37833/cord.v18i 01.355.

Singaravadivel, K., Alagusundaram, K., Hariharan, B. (2012). Physicochemical properties of fresh and stored coconut palm toddy. *Open Access Sci Reports* 1(8), https://doi.org/ 10.4172/scientificreports.397.

Sudha, R., Niral, V., Hebbar, K.B., Samsudeen, K. (2019b). Coconut inflorescence sap. *Current Sci* 116(11):1809–1817, https://www.jstor.org/stable/27138131.

Sudha, R., Niral, V., Hebbar, K.B., Samsudeen, K., Renjini, T.N.(2019a). Evaluation of coconut genotypes for coconut inflorescence sap yield. *Int J Tropical Agric* 37(2): 291-295.

Sudha, R., Niral, V., Samsudeen, K., Hebbar, K.B. (2021). An investigation of yield and quality of coconut inflorescence sap in different coconut (*Cocos nucifera* L.) genotypes under West Coast of India. *Int J Curr Microbiol App Sci.* 10(02): 3030-3041, doi: https://doi.org/10.20546/ijcmas.2021.1002.332.

Tulalo, A.M., Mawardi, S. (2018). Potential sap and coconut sugar production of three accession dwarf coconut. *Jurnal Littri* 24(2):87-92.

Wasantha, S.M., De Costa, W.A.J.M., Sangakkara, U.R., Jayasekara, C. (2009). Estimation of water use of mature coconut (*Cocos nucifera* L.) cultivars (CRIC 60 and CRIC 65) grown in the low country intermediate zone using the compensation heat pulse method (CHPM). *J National Sci Foundation Sri Lanka* 37(3): 175-186.

Ysidor, K.N., Rachel, A.R., Louis, J.K.K., Muriel, O.D.J., Prades, A., Kouassi, A., Marius BGH (2014) Glucide factors of the inflorescence sap of four coconut (*Cocos nucifera* l.) cultivars from Côte D'ivoire. *Int J Biochem Res Review* 4(2): 116-127, https://doi.org/10.9734/IJBCRR/2014/7148.

Chapter 3

Coconut Palm Sap Yield: The Effects of Environment, Irrigation and Nutritional Requirements

V. Krishnakumar*
ICAR-Central Plantation Crops Research Institute, Kasaragod, Kerala, India

Abstract

Coconut inflorescence sap, also known as *neera*, is a naturally sweet liquid extracted from the inflorescence of the coconut palm. It has attracted attention of consumers across the world due its high nutritive value, low glycemic index, agreeable flavor, and delicious taste. The coconut palms can be tapped when they attain yield stabilization, and tapping can be continued for the next two decades or so. The yield of coconut sap is influenced by various factors such as variety, health of palms, season and weather conditions during tapping, time of tapping, geographical location and soil fertility status, use of preservative substances, application of growth regulators, as well as tapper's skills. The coconut is a perennial palm and with initiation of flowering, it continues to produce at least one inflorescence every month and thus palms will have simultaneous vegetative and reproductive growth stages. Regular supply of nutrients and water has to be ensured for its sustained growth and production. To produce good quality coconut sap, meticulous planning of tapping and its collection processes are necessary. This chapter outlines aspects such as identification of palms, selection of inflorescence, time and frequency of tapping, seasonal impact on coconut

* Corresponding Author's Email: dr.krishnavkumar@gmail.com

In: Coconut Palm Sap
Editors: K. B. Hebbar, V. Krishnakumar and S. V. Ramesh
ISBN: 979-8-89530-640-6

sap yield, nutritional and water management for efficient and sustained sap production.

Keywords: coconut palm sap, palm nutrition, water requirement, irrigation schedule, sap yield

1. Introduction

Coconut inflorescence sap (*neera*) is a naturally sweet liquid extracted from the inflorescence of coconut palm. It has attracted attention due its high nutritive value, agreeable flavor, and delicious taste. *Neera* contains amino acids, minerals, vitamins, antioxidants, and provides many health benefits (Asghar et al., 2019; Leelayuwat et al., 2018; Zafar et al., 2019). Once extraction of sap commences, the palm can be tapped for the next 20 years. Banzon (1984) reported that the highest coconut sap yield in ethanol terms is 109 MJ month^{-1} palm^{-1} equivalent to oil yield from a 20-nut fruit bunch, thus indicating the maximum possible energy harvest from coconut sap than from coconut oil. Basu (2018) opined that from a sustainability point of view, tapping coconut trees for their sap (which yields 5,000 l ha^{-1}) provides 5–7 times higher harvestable energy (food production) ha^{-1} than coconut oil production from mature coconuts.

The unfermented coconut sap collected under controlled conditions has a slightly alkaline pH, golden-brown colour and mild sweetness. The fresh sap is oyster white in colour and translucent (Gupa et al., 1980). However, the traditional sap collected without any control of temperature leads to microbial fermentation, resulting in an acidic, off-white sap with a distinct fermented odor (Hebbar et al., 2015; Purnomo 2007). Microbial growth affects the quality of coconut sap during harvesting and transportation. Therefore, adoption of various hygienic practices during the entire tapping and collection process are necessary (Borse et al., 2007; Ghosh et al., 2018; Ysidor et al., 2015). Use of natural and synthetic preservatives prevents undesirable microorganisms (Adisetya et al., 2022; Raharjo et al., 2023; Saraiva et al., 2023; Somawiharja et al., 2018). The sap is extracted generally during early morning hours, and the tapping methods followed vary from one coconut growing country to another as well as within the same country.

2. Selection of Palms and Inflorescence for Tapping

According to Pethiyagoda (1979), though good sap yielding palms within the same cultivar could be identified, high nut yielding coconut palms may not necessarily be high sap yielders. Therefore, selection of palms for sap collection is very important. In order to extract sap, the inflorescence (spadix) is to be selected before it opens. Menon and Pandalai (1958) suggested that a spadix in a coconut palm becomes ready for tapping when the one immediately preceding it has just burst open or is about to do so. They also indicated that a swelling inside the spadix, due to button development, is considered an appropriate stage for tapping. Santiago and Virtudazo (1996), based on an elaborate survey at Leyte, Samar and Cebu, in the Philippines, reported that a high sap yielding tree has drooping leaves and has pliant leaflets. Nathanael (1966) also opined that palms with fairly shiny and pliable leaflets and whose spadix has a thin inner bract are ideal for tapping. Jeganathan (1974) indicated that coconut palms with longer leaves and leaflets are also suitable for sap production. Dwarf coconut palms producing tender nuts with large volume of water having good flavour and higher content of total and reducing sugars, as well as tall palms producing nuts with higher proportions of kernel (>30%) and husk (>40%) are suitable for tapping since such palms produce plenty of sap.

Coconut trees are usually tapped at an early age once they attain yield stabilization. The inflorescence chosen for tapping is tied around using a strong coir or plastic rope to prevent it from burst opening. Such selected inflorescence is prepared using a hammer, hand rubbed or massaged twice a day using palm of the tapper during morning and evening for about a week. After 4–5 days of such preparation, 7–10 cm of the tip of spadix is removed using a sharp knife. The sap begins to ooze out from the cut surface within a day or two and is collected in suitable containers.

3. Frequency of Tapping

The frequency of tapping varies according to the nature of palm, health of spadix, prevailing weather conditions, locality and soil fertility status (Pethiyagoda 1978). The sap yield is also influenced by the skill of tappers, as highly skilled tappers can tap the spadix for two months as against 30–45 days of average tapping.

In Sri Lanka, a period of 31 days of sap flow per spadix is the average where coconut palms can be tapped throughout the year as long as rainfall is satisfactory. Each palm is tapped for eight months followed by a resting period of four months (Coconut Research Institute 1967).

The tapping is usually done twice a day, both morning as well as evening. Each time, a length of 1–2 mm of spadix is removed using the special knife. A single spadix can be tapped until it reaches a length of about 10–15 cm from its base. Around three weeks before reaching this point, another spadix can be identified to continue the tapping and sap production (Nathanael 1970). Sarma et al., (2022) suggested that tapping can be undertaken throughout the year as coconut palm produces inflorescence regularly in the successive leaf axils.

4. Yield of Sap

Well-managed and high yielding palms produce a good quantity of sap. Wide variation exists from palm to palm for sap yield even within a variety. Tall palms yield more sap than dwarf palms. Though short stature and possibility of accommodating high density of palms per unit area makes dwarf palms ideal for tapping, fresh sap yield per palm is less since the size and length of spadix is short in most of the dwarf varieties as compared to tall and hybrids. Malaysian dwarf coconut varieties are generally healthier and therefore, yield more sap and are ideal for sap production. Secretaria et al., (2002) recorded the length of spadix and number of leaves on the crown as important traits having positive and significant correlation with sap yield. Other contributory factors affecting sap yield are the varietal differences in terms of vegetative and floral characteristics of coconut palms. Mathes (1984) reported that female flower production in palms which are tapped is much greater than in untapped palms, and hence, tapped palms can produce higher yield of mature nuts.

In Thailand, the cut inflorescence axis can secrete about 4 l of sap day^{-1} over a period of 60 days (Law et al., 2011). Under Tuvalu conditions, Hemstock (2013) found the average sap production per tree to range from 1.7 to 4.3 l day^{-1}. Under Indian conditions, a high yielding healthy tall coconut palm under good management has the potential to give sap yield of 50 l spadix^{-1} during 30 days of tapping. If six such healthy trees with long spadix are tapped for a period of 6 months, an average of 300 l of *neera* can be obtained in a year (Jnanadevan 2013). Yarin et al., (2019) suggested that instead of removing any unproductive palms, production of inflorescence sap from such palms can give a very good monetary return and ensure livelihood security of farmers.

On an average a spadix can produce 1.5–3 l of sap day^{-1} or 60–80 l in a span of 40–45 days. In such case, tapping only 50% of the 12 spadices produced in a palm and allowing the remaining to produce nuts can yield around 400 l of sap (Hebbar et al., 2015). Wijonarko et al., (2022) from Indonesia reported that weather conditions such as temperature and relative humidity (RH) are factors affecting the microbial profile, brix value, and total sugar content of coconut sap. The application of growth regulators like ethrel @ 2.5% has been shown to increase sap yield along with sugar content (Ranasinghe and Waidyanatha 2003).

5. Seasonal Impact and Sap Yield

Among the various factors that affect the quality of coconut sap, weather conditions (temperature, humidity, rainfall), heat/solar radiation, altitude, soil conditions (soil type and fertility) play a major role (Mustaufik et al., 2022). Menon and Pandalai (1958) noticed that sap flow increased during cloudy weather and gradually reduced during dry weather. The coconut palm can be tapped throughout the year since there is regular production of spadix in consecutive leaf axils. However, a tree yields maximum quantity of sap during certain seasons of the year. In India, tapping and collection of *neera* are mostly carried out in dry season from November to March and in wet season from April to October. Tapping during dry weather condition is performed mostly in low lying lands where palms do not suffer any moisture stress. From studies done at two experimental sites at Zamboanga City, the Philippines, Secretaria (1998) reported that besides the differences in coconut varieties, climatic conditions and nutrition status of palms contributed significantly to the differences in sap and nut yields. Santiago (1983) observed a negative relationship between temperature and sap yield.

In Sri Lanka, coconut palms could be tapped for many seasons as long as the rainfall is sufficient and palms are rested for four months in a year. In drier areas, however, the spathes produced in palms after 3–5 consecutive seasons of tapping diminish in size and eventually abort. Woodroof (1970) found that in one tapping day, the yield varied from 600 ml day^{-1} in dry areas to 2.8 l day^{-1} in wet districts. Tapping of palms during November yielded an average of 17 l, while it was 14 l palm^{-1} during April for an average tapping period of 31 days. Some dilution of *neera* was also observed during wet weather.

The weather conditions also influence the biochemical properties of coconut sap. According to Naknean et al., (2010), the composition and quality of sap depend on place, time, season and duration of tapping. During dry season, there will be the possibility of rapid growth of microorganisms. Lantemona et al., (2013) found from North Sulawesi Province, Indonesia that the volume of coconut sap tapped in rainy season was higher than that in dry season. The coconut sap obtained in heavy rain contained more water so that it reduced the total soluble solids (TSS) of sap. In addition, they also reported that in rainy season, the photosynthesis process cannot yield the maximum results due to less sunlight intensity.

From a study undertaken at the ICAR-Central Plantation Crops Research Institute (ICAR-CPCRI), Kasaragod, India, Samsudeen et al., (2013) obtained higher sap production of 1043 and 1003 ml palm^{-1} day^{-1} during post-monsoon (October-November) and monsoon (June-September) periods, respectively. Sap production was the lowest (899 ml palm^{-1} day^{-1}) during summer months (March-May), but was similar to that obtained (971 palm^{-1} day^{-1}) during winter (December-February). Rain influenced the sap production as observed from higher sap production during monsoon and post-monsoon periods. Sap flow increased only after middle of June indicating soil moisture build up due to onset of south west monsoon positively affecting the sap flow. Temperature during the period also had its impact on sap production as seen from the lowest production in summer. In general, sap production was negatively affected by increasing temperature. Wind velocity and evaporation were found to have significant negative correlation with sap production. The rate of sap collection also differed between day and night. Sap generally flows more during the evening hours (3:30-6:00 pm) than during the morning hours (7:30-10:00 am).

Haryanti et al., (2018) from central Java, Indonesia studied various chemical properties of coconut sap (pH, water content, TSS, reducing sugar, sucrose content, amino acid content, total polyphenol, vitamin C and antioxidant activity) tapped during three weather conditions, i.e., sunny day, drizzling rain, and heavy rain with an average relative humidity of 91.5, 92.5 and 93.5%, respectively. Variation of weather conditions during tapping did not affect pH value of coconut sap. The tapping in sunny day produced coconut sap with lower water content (85.41-86.41%), higher TSS (16.0-16.8 ^{o}brix), higher sucrose content (10.5-13.64 g 100g^{-1}), and least amount of reducing sugar (0.04-0.12 g 100 g^{-1}) compared to other two weather conditions. Tapping done either during a drizzling rainy day or heavy rainy period produced sap with high water content, and lower sucrose content. The sap produced during drizzling rain had the highest vitamin C content (66.58

mg 100 g^{-1}), and 66.67% antioxidant activity evaluated with 2,2-diphenyl-1-picrylhydrazyl (DPPH) radical scavenging assay (RSA). The sap produced during the heavy rainy period had the lowest free amino acids content (0.021g 100 g^{-1}).

Influence of five genotypes (Jamaica Tall, Neu Quewen Tall, Orissa Tall, Gangapani and Benaulim) and three seasons viz., monsoon (June to Sept.); post monsoon (Oct. to Nov.), and winter (Dec. to Feb.) on yield and quality of coconut inflorescence sap was studied at ICAR-CPCRI, Kasaragod, India by Sudha et al., (2019). The highest sap yield was recorded during post-monsoon (Oct.-Nov.) and monsoon (June-Sep.) seasons while the lowest yield was recorded during winter season (Dec.-Feb.) in all the genotypes. Winter season recorded the highest quality parameters (pH, TSS, total and reducing sugars, phenol and protein contents, the highest ascorbic and amino acids) in all the varieties.

Legaspi (2024) evaluated the performance of four dwarf coconut varieties (Aromatic-AGD; Equatorial Green Dwarf- EGD; San Isidro Green Dwarf-SGD, and Talisay- TAL) when tapped during two seasons (dry season: Jan.-Aug. and wet season: Aug.-Dec.) from a study done at the Capiz State University Burias Campus, the Philippines. The length of spadix was affected by season and the spadices were longer during dry season compared with those spadices during wet season. This observation was in agreement with the findings of Legaspi (2006). She further reported that spadices tapped during dry season were produced during wet season, where water is abundant, whereas spadices tapped in wet season were formed during dry season, where water is a limiting factor affecting the growth and development of palms.

Chan and Elevitch (2006) stated that between 12 and 15 spadices are produced at regular intervals; however, drought conditions may cause the spadix to stagnate or stop producing. The spadices tapped during wet season required a longer time compared to those spadices in dry season. AGD tapped during dry season had the longest tapping duration (40.34 days), which was comparable with the tapping duration of SGD (38.21 days) and EGD (31.55 days) during dry season and AGD, EGD and SGD during wet season tapping (around 28 days). The shortest tapping duration was recorded from TAL in both dry (12.78 days) and wet seasons (20.88 days). AGD tapped during the wet season produced the highest volume (ml) of sap spadix^{-1} (8821.33) followed by SGD (6155.69), and EGD (4785.7) tapped during wet season. The lowest amount of toddy was produced by TAL (1353 and 1664.22 ml during dry and wet seasons, respectively) which was comparable with volume of sap produced by the other three varieties during dry season. In general, more volume of sap was produced during wet season compared to dry season. This might be linked to the availability of water during the translocation process, when more water is available, greater the sap or exudates (Legaspi 2024).

The variety of coconut also influences the quantity and quality of inflorescence sap produced. The sap production potential of three spathes of ranks 7, 8 and 9 of two coconut varieties (Malayan Yellow Dwarf (MYD) and West African Tall (WAT)) and two hybrids (PB113+. i.e., CRD x improved RIT), PB121+ i.e., MYD x improved WAT) was studied at Côte d'Ivoire by Konan et al., (2013). For different cultivars, spathes of rank 8 produced the highest volumes of sap, ranging from 4.54 l in MYD to 61.81 l spathe^{-1} in PB 113+, while the lowest sap volumes were recorded from spathes of rank 9, varying from 0.26 l (MYD to 4.24 l (PB 113+). The total production of sap as well as duration of production differed among the cultivars also. The order for sap produced from the 8th spathe being PB 113+> PB121+>WAT>MYD, the respective quantities being 61.81, 36.43, 14.64, and 4.54 l spathe^{-1}; and durations being 46.78, 43.33, 37.78, and 24.0 days. That duration depended on length of spathes and regular sap flow allowed by them.

The sap production potential, tapping duration, sucrose content and pH of 7 different dwarf coconut accessions (Nias Yellow Dwarf, Raja Brown Dwarf, Salak Green Dwarf, Sagerat Orange Dwarf, Jombang Green Dwarf, Tebing Tinggi Dwarf, and Bali Yellow Dwarf) were evaluated in Indonesia by Mashud and Matana (2014). The overall average sap yield ranged from 0.83 to 2.17 l palm^{-1} day^{-1}, duration from 12.5 to 19.0 days spadix^{-1}, sucrose from 1.51 to 14.5%, and pH from 6.17 to 6.41. Tebing Tinggi Dwarf topped the table with average sap production (l palm^{-1} day^{-1}), duration (days spadix^{-1}), sucrose (%) and pH with 2.17;19.0; 14.5; and 6.41, respectively. The lowest values for all these parameters were shown by Raja Brown Dwarf (0.83; 15.0; 13.31; and 6.17, respectively). The variability of sensory acceptance and flavors of inflorescence sap from different cultivars of coconut grown in Côte d'Ivoire was studied by Ysidor et al., (2017). The physicochemical parameters in inflorescence saps of Malayan Yellow Dwarf, West African Tall, PB121+, and PB113+ coconut cultivars are presented in Table 1.

As the sap of PB113+ hybrid and WAT were more appreciated by consumers, the authors suggested them to be promoted for sap production as fresh drink, whereas sap produced from MYD and PB121+ could be processed into coconut by-products as vinegar, alcohol, syrup, or coconut sugar for further value addition.

Table 1. Physico- chemical parameters in the inflorescence saps of Malayan Yellow Dwarf, West African Tall, PB121+, and PB113+ coconut cultivars (Ysidor et al., 2017)

Parameters	Means (±standard deviation) per cultivar			
	MYD	WAT	PB 121+	PB 113+
pH	7.32 ± 0.17a	7.18 ± 0.22b	7.10 ± 0.2b	6.97 ± 0.15c
TTA (g acid eq/100g)	0.26 ± 0.04b	0.30 ± 0.08a	0.31 ± 0.06a	0.31 ± 0.05a
TCC (g/100 ml)	16.08 ± 0.75a	14.23 ± 0.76b	14.14 ± 1.1b	14.54 ± 1.14b
TAC (g/100 g)	0.26 ± 0.01b	0.23 ± 0.04c	0.23 ± 0.02c	0.29 ± 0.04a

TTA: Total titratable acidity; TCC: Total carbohydrates content; TAC: Total ash content.

6. Water Management and Sap Production

Water is needed for all metabolic and manufacturing activities in plants. Different land scenarios, soil types, climate conditions, seasons, and water supply situations are all factors that affect growth and yield of a crop. Effective management of available water resources for palms is very important. Large palms that are grown in warm, moist tropical environments are highly productive and support a large leaf area, and hence require large amounts of water. In the case of coconuts, production is highly influenced by weather-related changes than in other tree crops. The development of palms, production and harvest of nuts are frequently hampered by soil moisture stress caused by drought. Irrigation and fertilization, with a focus on the water cycle and fertilizer rotation, are critical to production and quality of coconut sap. Though results from many studies on the effect of irrigation on growth and productivity of coconut palms are available, not much research studies are conducted to understand the relationship between irrigation and inflorescence sap production in coconut.

The sap flow in coconut depends on variety and soil condition on which palm is grown. The diurnal variation and daily total water use of two coconut varieties (CRIC 60 and CRIC 65) growing on two contrasting land suitability classes (Kurunegala series, and Kuliyapitiya series) in the low country intermediate zone of Sri Lanka was studied by Madurapperuma et al., (2009). Diurnal variations of sap flow were observed on successive days, averaging about 3 l h^{-1} at night before and rising rapidly from about 06:00 h until 10:00 h. The sap flow remained high until about 16:00 h, which declined later on. The average daily total sap flow (l day^{-1}) of CRIC 60 and CRIC 65 in the Kurunegala series were 120 and 90, while it was 92 and 79, respectively in the

Kuliyapitiya series. Palms growing in the former series showed higher daily sap flow than those growing in the latter series with 27% and 15% increases in CRIC 60 and CRIC 65, respectively. The greater depth, and water holding capacity of soil, as well as greater trunk water storage of palms were the probable reasons for higher sap flow noticed in the Kurunegala series. Samsudeen et al., (2013) observed an increase in sap production in West Coast Tall (WCT) palms with irrigation (960 ml day^{-1}) when compared with palms under rain fed condition (460 ml day^{-1}) mostly due to higher water availability in irrigated soil.

7. Nutritional Management and Sap Production

Coconut is a perennial palm, and once it starts flowering, continues to produce at least one inflorescence every month and has simultaneous vegetative and reproductive growth stages. Therefore, palms need regular supply of nutrients for its sustained growth and production.

The effect of foliar application of different doses of liquid organic fertilizer (containing N, P, K, Ca, Mg and S) and intervals on *neera* production was studied in a coconut garden with sandy soil, located at the Kebumen Regency, Central Java, Indonesia by Purwanto et al., (2017). The concentration of foliar liquid organic fertilizer was adjusted at 24 ml l^{-1}. *Neera* production increased by 15.32% with foliar liquid organic fertilizer compared to control, and the highest production was obtained with 750 ml plant^{-1} and three weeks interval. From another study, Purwanto et al., (2018) showed that organic liquid fertilizer application in coconut foliage increased the coconut sap production (10.89%) and sugar brix (one point) in dry season.

The available nutrient content of soil and concentration in tissue influences the quantity and quality of coconut inflorescence sap. The requirement of nutrients of coconut palms that are tapped will be much higher than palms maintained for nut production. The effect of different levels of fertilizers on yield and biochemical properties of inflorescence sap was studied by Raghu and Joseph (2020) at the Instructional farm, College of Agriculture, Kerala, India. Fertilizers were applied at the recommended dose of fertilizers (RDF) and 25,50 and 75% more than the RDF to palms. Fully emerged unopened inflorescences were selected for tapping 11 months after treatment application. Application of fertilizers increased sap yield, the highest yield with 175% RDF (3.32 l day^{-1}) followed by 150% (2.78 l day^{-1}), 125% (2.32 l day^{-1}) and the least by 100% (1.42 l day^{-1}). The duration of sap production (45.4-48.2 days) did not differ due to different rates of fertilizer application. The pH differed due to fertilizer application, which showed a decreasing trend with increase in NPK fertilizer level, the highest value of 6.70 being recorded with 100% of RDF. The application of fertilizer also influenced various biochemical properties like total sugars, reducing and non-reducing sugars, phenol, and vitamin C, the highest values of 10.84, 0.57, 10.27 g 100 ml^{-1}, 2.56 and 2.19 mg 100 ml^{-1}, respectively being recorded with 175% dosage.

Matana et al., (2022) studied sap characteristics of six dwarf coconut varieties applied with different fertilizer doses during various seasons at the North Sulawesi, Indonesia. The quantity of sap produced by varieties followed the order: Tebing Tinggi Dwarf > Bali Yellow Dwarf> Salak Dwarf>Nias Yellow Dwarf> Raja Dwarf. Tebing Tinggi Dwarf and Nias Yellow Dwarf produced more sap with a fertilizer dose of urea 500 g + SP36 375 g + potassium chloride 750 g + kieserite 100 g, whereas Bali Yellow Dwarf and Salak Dwarf responded to the higher dose of urea 750 g + SP-36 562.5 g + potassium chloride 1125 g + Kieserite 150 g palm$^{-1.}$ The average sugar contents in the sap of dwarf varieties during the collections done in April, May, and June were 14.20-14.70%, 12.20-14.60%, and 12.87-14.38%, respectively. The highest yield of coconut sugar was observed in sap extracted from Nias Yellow Dwarf (17.29%).

8. Effect of Time and Duration of Tapping

The quality of coconut sap is influenced by harvesting time (morning or evening), environmental conditions, and harvesting methods. The production of sap is related to physical activities of coconut palms such as photosynthesis, respiration, transpiration, microbiological activities as well as micro-climate (weather) factors (Ta'lin 2013). Traditionally, coconut sap is collected during morning and/or evening by farmers. However, it takes anywhere from 6 to 24 h to collect the sap. In general, the farmers, based on their habit, take up tapping coconut palms for sap at morning and evening. Optimization of tapping time is very important to produce coconut sap with good volume and quality. In Indonesia, coconut fresh sap is collected in a container tied up on the inflorescence after cutting it and left for about 9-12 h.

Purnomo and Surjoseputro (2001) opined that the sap should be collected without delay if no preservative is added to the collection container, or otherwise the sap gets fermented after 15 h. Secretaria et al., (2003) suggested harvesting of coconut sap twice a day in the sequential coconut toddy and nut production (SCTNP) scheme of the PCA Davao Research Center, the Philippines or in the traditional full toddy tapping scheme during the morning before 8.00 am for the production of 'tuba' or vinegar, and afternoon not later than 3.00 pm for production of sap. This sap will be sweeter and suitable as fresh sap drink and for syrup/sugar making.

The tapping of inflorescence during the afternoon and collecting fresh coconut sap next day morning (about 12 h duration) with preservative gave fresh sap with 18.72% sucrose content (Purnomo and Mufida 2004). The night time temperature is usually lower than day time which can slightly retard the fermentation process in the sap. In Indonesia, tapping coconut sap is performed by farmers twice a day, namely morning tapping until evening (6.00 am- 4.00 pm) and afternoon tapping until morning (4.00 pm - 6.00 am). The 12-16 h duration causes decline of sap quality due to influence of microbes and weather parameters such as temperature, humidity and rainfall (Sulistyaningrum et al., 2015).

The chemical properties of coconut sap collected from Dalam variety at different tapping time (twice a day in the daytime for 9 h from 6.00 pm–3.00 am, and at night time for 15 h from 3.00 pm - 6.00 am) and addition of lime and mangosteen peel powder as preservatives was studied by Haryanti et al., (2017) at the Sumbang District, Central Java, Indonesia. The tapping time and preservative affected chemical properties of coconut sap and palm sugar yielded. Due to the higher sucrose content and lower reducing sugar, the day time tapping was found to be ideal than night time tapping to produce granulated coconut sugar.

The duration of sap production or exploitation length of spadix is the number of days during which it produces sap. Konan et al., (2013) opined that the duration of sap production of coconut is related to the length of spadix and regular flow of sap during their exploitation. They also suggested that the volume of sap available is closely related to length of production and cultivar's nut yield. The time and duration of tapping exert significant changes in the quality of coconut sap. If the tapping time is too long (i.e., >8 h), it will cause a decrease in the quality of coconut sap, especially pH and sucrose content due to fermentation process of coconut sap. The perishable nature of coconut sap will also result in variation of quality of palm sugar produced.

In the Banyumas Regency of Central Java Province, Indonesia, coconut sap tapping is typically done twice a day, i.e., during daytime for about 9 h and during night for about 15 h. Haryanti et al., (2017) studied the effect of sap tapping during night and day time along with use of preservatives such as lime and mangosteen peel powder. The difference of tapping time and preservative significantly affected various chemical properties of coconut sap and palm sugar yield. Though tapping during night time showed lower values for pH, TSS and sucrose content in the sap, the water content and reducing sugar were found significantly higher than tapping of sap during day time. Since day time tapping resulted in higher sucrose content and lower reducing sugar, the sap was suitable as raw material to produce granulated coconut sugar. The optimum concentration of the preservatives added was 0.56 g l^{-1}.

Mustaufik et al., (2021) evaluated the effect of tapping time, duration of tapping, and usage of natural laru (a mixture of lime with mangosteen rind extract and betel leaf) as preservative on quality of coconut sap (volume, discharge, acidity level/pH and sucrose content) at Central Java, Indonesia using 15-20 years old coconut palms of Dalam variety. The tapping done from evening to night (14.00 - 22.00 pm) gave sap yield of 511.32 ml, which was higher than that produced (466.34 ml) from tapping done during morning to noon (6.00 am-14.00 pm). The degradation of physicochemical quality of coconut sap (especially pH and sucrose content) was slower in tapping during evening to night than during the other time of tapping. The optimal limit of tapping time with and without laru during morning to noon was 8 and 6 h, respectively, while for both the cases, it was 8 h for the evening to night tapping. The increase in tapping time may increase the concentration of reducing sugars due to microbial activity.

Priya et al., (2023) collected inflorescence sap produced by traditional tapping method (i.e., with lime coating inside the pot) at regular intervals during overnight and day time and analysed for various physicochemical quality attributes (pH, TSS, titratable acidity, reducing and total sugars, protein content, total phenols, alcohol content, color, and turbidity). Collection of inflorescence sap during morning hours (6.00 am –9.00 am) resulted in optimal quality of sap. The sap without lime addition increased the acidity, reducing sugars, microbial population and decreased the total sugars and protein content resulting in reduced shelf life of sap. The sap tapped at 9.00 am with lime addition provided desirable conditions that enabled maintenance of physicochemical properties and product quality at an optimum level.

Conclusion

Coconut is a perennial palm, and once a palm starts flowering, it continues to produce at least one inflorescence every month and thereafter undergoes simultaneous vegetative and reproductive growth stages. Therefore, the palm needs regular supply of nutrients and water for its sustained growth and production. In order to ensure the quality of coconut sap produced, meticulous planning of tapping and its collection processes are necessary.

References

Adisetya, E., Krisdiarto, A.W. (2022) Preservative of coconut sap shelf life derived from mangosteen yellow latex. *Jurnal Ilmiah Teknologi Dan Industri Pangan Unisri* 7(1): 59–67, https://doi.org/10.33061/jitipari.v7i1.6757.

Asghar, M.T., Yusof, Y.A., Mokhtar, M.N., Ya'acob, M.E., Mohd. Ghazali, H., Chang, L.S., Manaf, Y.N. (2019) Coconut (*Cocos nucifera* L.) sap as a potential source of sugar: Antioxidant and nutritional properties. *Food Sci Nutr* 8(4):1777–1787, doi:10.1002/fsn3.1191.

Banzon, J.A. (1984) Harvestable energy from the coconut palm. *Energy Agric* 3: 337-344, https://doi.org/10.1016/0167-5826(84)90044-5.

Basu, M.S. (2018) Coconut tree: tap for sap to sugar. ttps://www.linkedin.com/pulse/coconut-tree-tap-sap-sugar.

Borse, B.B., Rao, L.J., Ramalakshmi, M.K., Raghavan, B. (2007). Chemical composition of volatiles from coconut sap (neera) and effect of processing. *Food Chemistry* 101 (3):877–880, doi:10.1016/j.foodchem.2006.02.026.

Chan, E., Elevitch, C.R. (2006) *Cocos nucifera* (coconut). Species Profiles for Pacific Island Agroforestry. *Permanent Agriculture Resources*. 27p.

Coconut Research Institute (1967) Toddy tapping. Ceylon. *Leaflet* No. 48. 13p.

Ghosh, D.K., Bandyopadhyay, A., Das, S., Hebbar, K.B., Biswas, B. (2018) Coconut sap (Neera)-untapped opportunity of spin off gains in West Bengal, India. *Int J Curr Microbiol Appl Sci* 7 (7):1883–1897, https://doi.org/10.20546/ijcmas.2018.709.229.

Gupta, R.C., Jain, V. K., Shanker, G. (1980). Palm sap as a potential starting material for vinegar production. *Research Industry* 25: 5-7.

Haryanti, P., Supriyadi, Marseno, D.W, Santoso, U. (2018) Effects of different weather conditions and addition of mangosteen peel powder on chemical properties and antioxidant activity of coconut sap. *Agritech* 38 (3):295-303, https://doi.org/10.22146/agritech.29844.

Haryanti, P., Supriyadi, Marseno, D.W., Santoso, U. (2017) Chemical properties of coconut sap obtained at different tapping time and addition of preservatives. The Int J Sci Technol 5(3): 52-59.

Hebbar, K.B., Arivalagan, M., Manikantan, M.R., Mathew, A.C., Thamban, C., Thomas, G., Chowdappa, P. (2015) Coconut inflorescence sap and its value addition as sugar -

collection techniques, yield, properties and market perspective. *Curr Sci* 109:411–1417, https://doi.org/10.18520/cs/v109/i8/1411-1417.

Hemstock, S.L. (2013)The potential of coconut toddy for use as a feedstock for bioethanol production in Tuvalu. *Biomass and Bioenergy* 49:323–332, doi: 10.1016/j.biombioe.2012.09.005.

Jeganathan, M. (1974) Toddy yield from hybrid coconut palms. *Ceylon Coconut Quarterly* 25:139- 148.

Jnanadevan, (2013) Coconut palms suitable for neera tapping. *Indian Coconut J* 56(1):15–16.

Konan, N.Y., Konan, K.J.L., Assa, R.R., Konan, B.R., Okoma, D.M.J., Allou, K., Biego, G.M.H. (2013) Assessment of sap production parameters from spathes of four coconut (*Cocos nucifera* L.) cultivars in Côte d'Ivoire. *Sustain Agric Res* 2(4):87-94.

Lantemona, H., Abadi, A.L., Rachmansyah, A., Pontoh, J. (2013) Impact of altitude and seasons to volume brix content, and chemical composition of aren sap in North Sulawesi. *IOSR J Environ Sci Toxicol Food Technol* 4(2):42–48, doi:10.9790/2402-0424248.

Law, S.V., Bakar, A.F., Hashim, M.D., Hamid, A.A. (2011) Popular fermented foods and beverages in Southeast Asia. *Int Food Res J*18(2):475–84.

Leelayuwat, N., Saelim, B., Kanpettha, Y., Aneknan, P.T., Tong, U.N.T. (2018) Effects of organic coconut flower syrup on glucose and insulin responses and substrate utilization. *Asia Pac J Sci Technol* 23 (4): 1–6, https://doi.org/10.14456/apst.2018.16.

Legaspi. S.J. (2006) Yield and quality of toddy from tall and dwarf varieties of coconut (*Cocos nucifera* L.) as influenced by inflorescence morpho-anatomy and season. University of the Philippines Los Baños (UPLB). Doctoral Thesis Dissertation.

Legaspi. S.J. (2024).Floral attributes and toddy yield of selected dwarf coconut varieties as affected by tapping season. *Sustainable Energy Environ Review* 2(1):1-7,doi: 10.59762/seer924712041120240705093415.

Madurapperuma, W.S., de Costa, W.A.J.M., Sangakkara, U.R., Jayasekara, C. (2009) Estimation of water use of mature coconut (*Cocos nucifera* L.) cultivars (CRIC 60 and CRIC 65) grown in the low country intermediate zone using the compensation heat pulse method (CHPM). *J National Sci Foundation Sri Lanka* 37 (3): 175-186, doi: 10.4038/jnsfsr.v37i3.1211.

Mashud, N., Matana, Y.R. (2014). Produksi nira beberapa aksesi kelapa Genjah. *Buletin Palma* 15(2): 110-114.

Matana, Y.R., Rindengan, B., Novarianto, H., Tulalo, M., Manaroinsong, E., Kumaunang, J. (2022) The effect of fertilizer to production of neera dwarf coconut. IOP Conf. Series: Earth and Environmental Science 974, 012093, doi:10.1088/1755-1315/974/1/012093, 6 pages.

Mathes, D.T. (1984) Effect of extraction of inflorescence sap and the removal of immature fruit bunches on the production of female flowers in coconut. *COCOS* 2: 44-47.

Menon, K.P.V., Pandalai, K.M. (1958) *The Coconut Palm – A Monograph*, Indian Central Coconut Committee 384p.

Mustaufik, Sutiarso L., Rahayoe, S., Widodo, K.H. (2021) The effect of time and duration of tapping and the addition of laru as natural preservative in coconut sap quality. IOP

Conf. Series: Earth and Environmental Science 653,doi:10.1088/1755-1315/653/1/012084, 10 pages.

Mustaufik, Sutiarso, L., Rahayu, S., Widodo, K.H. (2022) Technique engineering of tapping and shelter of coconut sap and its effect on the quality of crystal coconut sugar. *Food Research* 6 (2) : 248 – 254, doi: https://doi.org/10.26656/fr.2017.6(2).220.

Naknean, P., Meenune, M., Roudaut, G. (2010). Characterization of palm sap harvested in Songkhla province, Southern Thailand. *Int Food Res J* 17(4):977–986.

Nathanael, W.R.N. (1966) A tapping technique for the coconut palm. *Ceylon Coconut Center's Review* 4(4):87- 99.

Nathanael, W.R.N. (1970) Coconut toddy tapping and cottage manufacture of treacle and jaggery. *Ceylon Coconut Center's Review* 6: 63-65.

Pethiyagoda, U. (1979) Coconut inflorescence sap. *Planter* 55(641):390–397.

Pethiyagoda, U. (1978) Coconut inflorescence sap. *Perak Planters' Association Journal.* Malaysia. pp 86-91.

Priya, V.V., Balakrishnan, M., Karthikeyan, S., Rajkumar, P., Manikantan. M.R., Gurusamy, K., Pandiselvam, R. (2023) Evaluation of effect of tapping frequency and lime coating on the biochemical changes and fermentation rate of coconut inflorescence sap. *Sugar Tech* 25: 906–915, https://doi.org/10.1007/s12355-023-01254-3.

Purnomo, H. (2007) Volatile components of coconut fresh sap, sap syrup and coconut sugar. *ASEAN Food J* 14(1):45-49, http://ifrj.upm.edu.my/afjv14(1)2007/45-49.

Purwanto, Mujiono, Tarjoko (2017) Effect of foliar liquid organic fertilizer on neera production. *Planta Tropika: Jurnal Agrosains* 5(2):106-109, doi: 10.18196/pt.2017.070.
106-109.

Purwanto, Tarjoko, Mujiono, and Pramono TB(2018) Increasing coconut sap production with liquid organic fertilizer application in dry season. *J Innov Applied Technol* 4(2):762-765.

Raghu, R.S., Joseph, B. (2020) Effect of different levels of NPK fertilizers on yield, and bioche,ical properties of inflorescence sap of coconut. *Int J Chem Stud* 8(3): 2126-2129.

Raharjo, D., Zaman, M.Z., Praseptiangga, D., Yunus, A. (2023) Physicochemical and microbiological characteristics of various stem bark extracts of *Hopea beccariana* Burck potential as natural preservatives of coconut sap. *Open Agriculture* 8: 20220175, https://doi.org/10.1515/opag-2022-0175, 19 pages.

Ranasinghe, C.S., Waidyanatha, U.P.S. (2003) Ethrel stimulation of inflorescence sap flow in tapped coconut (*Cocos nucifera*) palms. *Exp Agric* 39(2):161-166

Samsudeen, K., Niral, V., Jerard, B.A., Berwal, M.K., Sugatha, P., Hebbar, K.B. (2013) Influence of variety and season in the inflorescence sap production in *Cocos nucifera* L. *J Plantn Crops* 47(1): 57–61

Santiago, R.M. (1983) Toddy yield of coconut as influenced by fertilization, leaf nutrients, water status, climatic factors arid morphological characters. Ph.D. Dissertation UP at Los Banos, College, Laguna, Phil. 121 p.

Santiago, R.M., Virtudazo, M.S.J. (1996) Coconut tapping in the Philippines enigmas and observations. *CORD* 12(1):34-45, https://doi.org/10.37833/cord.v12i01.296.

Saraiva, A., Carrascosa, C., Ramos, F., Raheem, D., Lopes, M., Raposo, A. (2023) Coconut sugar: chemical analysis and nutritional profile; health impacts; safety and quality control; food industry applications. *Int J Environ Res Public Health* 20(4):3671, doi: 10.3390/ijerph20043671.

Sarma, C., Mummaleti, G., Sivanandham, V., Kalakandan, S., Rawson, A., Anandharaj, A. (2022) Anthology of palm sap: The global status, nutritional composition, health benefits & value added products. *Trends Food Sci Technol* 119: 530-49, http://dx.doi.org/10.1016/j.tifs.2021.12.002.

Secretaria, M.I., Ebuna, R.M., Madrazo, R.M., Magat, S.S., Maravilla, J.N., Santos, G.A, Baylon, G.B. (1998) Sequential coconut toddy-nut production (SCTNP) scheme in different coconut cultivars and hybrids: production and processing of product. *Philippine J Crop Sci* 23 (1): 76.

Secretaria, M.I., Ebuna, R.M., Magat, S.S. (2003) On-farm production and processing of selected coconut sap-based natural and nutritious food products from SCTNP scheme. *CORD* 19(2) doi: https://doi.org/10.37833/cord.v19i02.375, 12 pages.

Secretaria, M.I., Ebuna, R.M., Magat, S.S., Maravilla, J.N., Santos, G.A., Baylon, G.B. (2002) Performance of selected coconut varieties/hybrids under sequential coconut toddy and nut production scheme or SCTNP. *CORD* 18(1): 12-27, doi:10.37833/cord.v18i01.355.

Somawiharja, Y., Wonohadidjojo, D.M., Kartikawati, M., Suniati, F.R.T., Purnomo, H. (2018). Indigenous technology of tapping, collecting and processing of coconut (*Cocos Nucifera*) sap and its quality in Blitar Regency, East Java, Indonesia. *Food Research* 2(4) 1-6, https://doi.org/10.26656/fr.2017.2(4).075.

Sudha, R., Niral, V., Hebbar, K.B., Selvamani, V., Samsudeen, K. (2019) Influence of genotype and season on yield and quality of coconut inflorescence sap. *Int J Curr Microbiol App Sci* 8(6): 2280-2288, https://doi.org/10.20546/ijcmas.2019.806.270.

Sulistyaningrum, A., Yanto, T., Naufalin, R. (2015) Perubahan kualitas nira kelapa akibat penambahan pengawet alami. *J. Penelitian Pascapanen Pertanian, 12*(3), pp.137-146.

Tamrin, T. and Ta'lin, T.L., 2013. Modifikasi Dan Perbaikan Kinerja Alat Pengiris Mayang Kelapa Untuk Menyadap Nira. *Agritech*, *33*(4), pp.477-482.

Wijonarko, G., Purbowati, I.S.M., Widjojoko, T., Maksum, A. (2022) PengaruhagroklimatdesaGandatapakecamatanSumbangterhadapprofilmikrobia, nilai brix dan gula total nirakelapaberlarualami (The influence of agroclimate, Gandatapa village, Sumbang Subdistrict on microbial profile, brix values, and total sugar, naturally spread coconut sap). *Indonesian J Food Technol* 1(2): 41-54.

Woodroof, J.G. (1970) Coconuts: Production, Processing Products, The AVI publishing Company, Inc., Westport, Connecticut 241 p.

Yarin, T., Bandyopadhyay, A., Ghosh, D.K. (2019) Sustainable approach towards conversion of unproductive coconut palms into productive ones through neera production. *J Crop and Weed* 15(3): 12-18, doi: https://dx.doi.org/10.22271/09746315.2019.v15.i3.1229.

Ysidor, K.N.G., Louis, J.K.K., Rachel, A.R., Marius, B.G.H. (2017). Variability of sensory acceptance and flavors of the inflorescence sap deriving from four widespread cultivars of coconut (*Cocos nucifera* L.). *Current J Applied Sci Technol* 22(3):1-10. doi: 10.9734/CJAST/2017/33059.

Ysidor, K.NG., Louis, J.K.K., Roger, K.B., Rachel, A.R., Joelle, O.D.M., Emmanuel, I.A., Marius, B.G.H. (2015) Changes in physicochemical parameters during storage of the inflorescence sap derived from four coconut (*Cocos nucifera* L.) varieties in Coˆte d'Ivoire. *J Exp Agric Int* 352–365, https://doi.org/10.9734/AJEA/2015/11035.

Zafar, M.A., Mills, K.R., Zheng, J., Regmi, A., Hu, S., Gou, L., Chen, L, (2019) Low-glycemic index diets as an intervention for diabetes: a systematic review and meta-analysis. *Am J Clin Nutr* 110:891–902, https://doi.org/10.1093/ajcn/nqz149.

Chapter 4

Coconut Inflorescence Sap: Biochemical Composition and Quality Attributes

S. V. Ramesh[1,*]
V. Krishnakumar[2]
J. Anjitha[1]
S. Sudharshana[1]
and K. B. Hebbar[1,†]
[1]ICAR-Central Plantation Crops Research Institute, Kasaragod, Kerala, India
[2]ICAR-Central Plantation Crops Research Institute, Kayamkulam, Kerala, India

Abstract

Coconut inflorescence sap, commonly known as toddy or *neera*, is a natural liquid extracted from the unopened spadix of the coconut palm. This chapter delves into the biochemical composition and quality attributes of coconut inflorescence sap, highlighting its nutritional value and potential health benefits. The sap is rich in sugars, vitamins, minerals, amino acids, and antioxidants, making it a highly nutritious and functional beverage. The chapter discusses the genotypic influence on production of coconut inflorescence sap and various value-added products made out of it. Understanding the biochemical properties and quality characteristics of coconut inflorescence sap is crucial for optimizing its utilization and ensuring product safety and consistency.

---

[*] Corresponding Author's Email: ramesh.sv@icar.gov.in
[†] Corresponding Author's Email: hebbar.kb@icar.gov.in

In: Coconut Palm Sap
Editors: K. B. Hebbar, V. Krishnakumar and S. V. Ramesh
ISBN: 979-8-89530-640-6

Keywords: bioactives, nutritional composition, quality standards, phytochemicals

1. Introduction

Palms are monocotyledonous angiosperms belonging to the family Arecaceae, which includes 2,500 to 2,700 known species spread across six sub-families and 200 genera. Economically significant palms include *Cocos nucifera* (coconut palm), *Arenga pinnata* (sugar palm), *Hyphaene* spp. (real fan palm, ivory palm), *Phoenix sylvestris* (sugar date palm), and *Areca catechu* (betel nut palm). They have diverse applications in various industries, including coir, oil, sago, and sugar production. The young inflorescences of these palms are harvested to produce a variety of products, including fresh juice, wine, syrup, sugar, and jaggery (Hebbar et al., 2015; Kurniawan and Kustiningsih 2019;Nguyen et al., 2016).

Coconut sap, also known as *neera*, is a beverage traditionally tapped from the coconut spadix and widely consumed by rural populations. This phloem sap is rich in sugars, minerals, antioxidants, and vitamins, which the palm uses for the growth and development of tender or mature coconuts. The sap flows slowly and is highly prone to fermentation, making the collection of unfermented sap a challenging task. Lime is commonly used during tapping to inhibit fermentation, however, the sap often turns white and develops a foul smell, rendering it unsuitable as a health drink. The fermented sap, known as toddy, has a strong odour that makes it unpalatable despite its nutritional value (Hebbar et al.2015; Hebbar et al., 2020; Sudha et al., 2019).

The glycemic index (GI) and insulinemic index (II) categorize foods based on the levels of blood glucose and insulin they generate, which are linked to the risk of developing metabolic diseases. Zafar et al., (2019) proposed that low-GI diets can lower postprandial blood sugar, leading to more stable glucose levels. These diets may be beneficial for managing blood sugar and could potentially help reduce body weight in individuals with pre-diabetes or diabetes. Sap is classified as a food with a low GI (GI 35) (Trinidad et al., 2015). However, another group of researchers reported a GI of 52.17 for coconut sap, classifying it in the intermediate range (Chinnamma et al., 2019). Leelayuwat et al., (2018) from Thailand analyzed the GI and insulinemic index (II) of organic coconut sap and discovered that both values to be low (51.2 ± 9.26 for GI and 45.4 ± 10.0 for II. This suggests that coconut sap could be a suitable alternative to sugar as a sweetener, an important finding for preventing and reducing the risk of type 2 diabetes and other metabolic diseases.

It is assumed that variations in the GI of coconut sap could be attributed to factors such as the age of the tree, its variety, climate, and geographical location. Therefore, a comprehensive study is necessary to accurately determine the GI of palm saps. The most attractive aspect of the sap is its flavour and nutritional content. This organic sap is known for its numerous health benefits, including anti-diabetic, anti-aging, anti-rheumatic, and anti-electrolyte imbalance properties (Hebbar et al., 2015).

Despite its high nutritional value and health benefits, there is a lack of awareness about the sap among the general population. It is primarily consumed as a traditional drink in rural areas. Due to inadequate preservation techniques that hinder transportation, urban populations are generally unaware of this healthy drink. The sap's naturally high microflora content necessitates careful preservation. Additionally, the absence of technology for hygienic tapping under controlled conditions is a significant drawback, limiting the sap's shelf life (Chinnamma et al., 2019).

Implementing preventive measures in handling the sap during harvest, processing, and storage are strongly recommended due to its high susceptibility to instant fermentation. The rapid fermentation of the sap poses a significant challenge for preservation, which is why it is often converted into value-added products and marketed globally on a commercial scale. This chapter outlines and discusses the composition and quality attributes of coconut palm sap as well as production of value-added products, preservation methods, and health benefits.

2. Coconut Palm Sap

Fresh coconut inflorescence sap, obtained from the phloem sap of the unopened spadix of the coconut palm (Gopal et al., 2021), is sweet, oyster white, and translucent. It is recognized for its high nutritional content and its role as an effective digestive aid (Hebbar et al., 2018). Coconut inflorescence sap is the most profitable product for farmers among all coconut derivatives. It can be transformed into coconut palm sugar, jaggery, syrup, and other products, which are primarily consumed by rural communities (Priya et al., 2023). A mature coconut palm produces 12–14 inflorescences annually, roughly one per month (Divya et al., 2019). It is estimated that a healthy tree can yield 1.5–2.5 l of sap daily (Priya et al., 2023).

Palm sap is harvested by cutting the tip of the inflorescence. In rural areas, palm sap is traditionally harvested from palmyrah and coconut trees through an established practice for local consumers (Nathanael 1966). The inflorescence is tapped at a mature stage (Borse et al., 2007;Redhead 1989), identified by the swelling at its base, which indicates the right time for tapping. Sap flow is stimulated by gently and uniformly beating coconut inflorescences with traditional tools (bones or wooden sticks) every morning and evening for seven days. To prevent the inflorescence from bursting open, it is tied with strong coconut fiber or coir rope. Then, 7–10 cm of the front portion is cut off with a sharp sickle. In the traditional method, the tapped inflorescence is placed into an earthenware pot for sap collection. As the sap drips over time, it ferments in the open container and becomes contaminated with pollen, ants, insects, and other pollutants (Hebbar et al., 2018).

Table 1. Biochemical features of fresh sap and fermented sap (Sharma and Meghwal 2021)

Parameters	Fresh coconut inflorescence sap (FCIS)	Naturally fermented coconut inflorescence sap (NCIS)
pH	7-7.4	3.3-5.9
Acidity (mg l^{-1})	15.0 ± 0.8	18.0 ± 0.5 -177.0 ± 9.1
Phenolic content (μgl^{-1})		
Gallic acid	350 ± 1.36	500 ± 1.51
1. Protocatechuric acid	86 ± 0.37	210 ± 1.02
2. Caffeic acid	56 ± 0.21	730 ± 1.29
3. P- coumaric acid	27 ± 0.13	150 ± 0.98
4. Galangin	100 ± 0.65	120 ± 1.07
Colour	Oyster white	Cloudy whitish
Flavour compound		
Amino acid content	High	Low
Taste	Sweet	Sour, cider like taste
Shelf life	Processed – up to 6 months	Processed -1 day
Aromatic alcohols (μgl^{-1})	0	196.20
Phenolic compounds (μgl^{-1})	145.38	0
Aliphatic hydrocarbons (μgl^{-1})	411.326	0
Esters (μgl^{-1})	20.84	1032.76
Carbonyl compounds (μgl^{-1})	4.12	2249.65
Sucrose content (%)	14.35	1.04
Glucose content (%)	0.28	8.72
Fructose content (%)	0.32	5.75

In India, the inner surface of the earthen pots used for collecting coconut inflorescence sap is coated with lime to prevent fermentation, although this method only partially inhibits the process. In Thailand, Kiam wood (*Cotylelobium melanoxylon* Pierre) is placed in bamboo tubes during collection to slow down microbial growth (Naknean et al., 2014). The ICAR-CPCRI has developed the coco-sap chiller to collect hygienic (contaminant-free) and non-fermented sap which is called '*Kalparasa.*' A hollow PVC pipe is used to make it, and a bag is inserted at its top to collect sap. With the addition of ice cubes, the collection box maintains an internal temperature of 2–3°C for 10–12 h to prevent fermentation of the sap. The collected sap is therefore fresh, unfermented, hygienic, and delicious. Additionally, it has the potential to be processed into various value-added products such as syrup, squash, concentrate, jaggery, and sugar (Hebbar et al., 2018).

Palm sap, being phloem sap, is highly nutritious. Its composition and quality vary depending on the location and duration of tapping (Borse et al., 2007). As already stated, the sap exhibits a low GI of 35, contrasting with the higher GI of 65 found in refined white sugars (Trinidad et al., 2010). The freshly harvested sap is abundant in phenolics, vitamins (including B vitamins and ascorbic acid), protein, and essential elements such as nitrogen, phosphorus, potassium, and magnesium, as well as micronutrients like zinc, iron, and copper. Due to high potassium content, it aids in regulating high blood pressure and sugar metabolism. Additionally, it serves as a potent antioxidant, supporting the immune system, respiratory system, cardiovascular health, and reducing inflammation. Furthermore, it contributes to the electrolyte balance and maintains the acid/alkaline ratio (Hebbar et al.2015). The most attractive aspect of the sap is its flavour and nutritional content. The details of biochemical features including phenolic content of fresh sap and fermented sap (Sharma and Meghwal 2021) are presented in Table 1.

3. Genotypic Influence on Coconut Inflorescence Sap

Understanding the effects of coconut variety and seasonal influence on inflorescence sap production is vital for farmers. This knowledge helps select suitable varieties, regulate tapping schedules, and implement irrigation practices to boost sap yield. Tall coconut varieties are more effective than dwarf ones in producing inflorescence sap. Among the tested varieties, WCT and DxT hybrid had the highest sap yields, averaging 960 ml and 880 ml, respectively. In contrast, COD had very low sap production, making it unsuitable for tapping. WCT also had the highest total sugar content at 11.44 g 100 ml^{-1}, while DxT had a high reducing sugar content of 0.37 g 100 ml^{-1}. The ideal tapping period is from mid-monsoon to early winter, and tapping during summer, especially under rainfed conditions, should be avoided (Samsudeen et al., 2013).

Ysidor et al., (2014) suggested that the high glucose content in coconut inflorescence sap is linked to its total soluble solids (TSS), dry matter, and overall carbohydrate levels. The sap of the cultivars (MYD, WAT, PB121+, and PB113+) had a neutral pH, with the MYD cultivar contained the highest concentration of these macromolecules. Sucrose was identified as the primary carbohydrate, and its higher content in MYD made the sap particularly flavourful for drinking, while the hybrids (PB121+ and PB113+) showed intermediate values. Additionally, similar levels of glucose and fructose across the cultivars indicated that these monosaccharides were at the same synthesis stage. Therefore, the sap from these cultivars could yield comparable amounts of alcohol, syrup, and sugar, due to their high sucrose content.

A study in the Philippines on the Sequential Coconut Toddy and Nut Production (SCTNP) scheme across six coconut hybrids and cultivars in Zamboanga City found significant differences in sap and nut yields. D x T hybrids and the local tall cultivar, Laguna, had notably higher yields compared to the Catigan dwarf, Catigan dwarf hybrid (CAT x CAT), and a dwarf cultivar (Tacunan). Besides varietal differences, climatic conditions and the nutritional status of the palms greatly influencesap and nut yields (Secretaria 1998). Additionally, sap flow in coconut is affected by the variety's leaf area and the soil conditions, including soil depth and water-holding capacity, where the palm is grown.

4. Nutrition Profiling of Coconut Palm Sap

The composition and quality of fresh coconut sap are influenced by various factors, including the geographical region, coconut variety, maturity stage of the inflorescence, seasonal and climatic conditions, soil fertility, and the duration of tapping (Borse et al., 2007; Hori et al., 2001; Naknean et al., 2010).

Coconut treacle made from coconut inflorescence juice or the sap is a sugar-rich food with almost 65-75% sugars and 25-35% water and other minor constituents such as organic acids, mineral salts, vitamins, proteins, phenolics and free amino acids (Ouchemoukh et al., 2007). Coconut sap contains 0.26%

protein and 6.91% sugar (Azlan et al., 2020). Coconut palms have been found to contain higher levels of amino acids such as glutamic acid, serine, aspartic acid, and threonine. Table 2 and 3 provide details on the nutritional composition of coconut palm sap. The sap primarily consists of about 80% water and 10–15% total sugars, with sucrose being the main component.

Coconut sap from Malang, East Java was found to contain 70.85% sucrose, 3.00% glucose, and 2.92% fructose (Purnomo, 1992). Coconut inflorescence sap is typically neutral (pH 6.97 to 7.32) and rich in carbohydrates, with a content of 10-15%, primarily sucrose, similar to that of sugarcane (Nakamura et al., 2004). The sap's sugar content was measured at 339±47.2 mM sucrose, 9.6±7.8 mM fructose, and 5.5±1.2 mM glucose. Sucrose constitutes more than 80% of the sap's total solids, followed by small amounts of glucose and fructose, around 2.3% of total solids (Nurhadi et al., 2018). Asghar et al., (2020a) reported sugar contents of 6.91% sucrose, 3.48% fructose, and 2.53% glucose. Additionally, palm sap contains a significant amount of inulin, a fructose polymer that serves as a natural substitute for sugar and possesses prebiotic properties (Sudha et al., 2019). Fresh sap contains approximately 0.33 gl^{-1} of total phenolics (Xia et al., 2011).

The pH of fresh sap tends to be close to neutral, ranging from 6.6 to 6.8, while fermented sap typically exhibits a pH range of 3.5 to 4.0. Besides, it contains a variety of phytochemicals and volatile compounds, including aromatic hydrocarbons, aliphatic ketones, acids, alcohols, heterocyclic compounds, and esters. Among these volatiles, acetic acid and 2-butanol were identified as the primary compounds (Purnomo 2007).

Table 2. Nutritional composition of coconut palm sap (100 ml) (Hebbar et al., 2018)

Nutritional composition	Content
pH	6.5-7.5
Total sugar (g)	10.08-16.50
Total Reducing Sugar (g)	0.55
Sucrose (g)	13.40
Glucose (g)	1.50
Fructose (g)	0.60
Protein (g)	0.15-0.28
Total Phenols (mg gallic acid equivalent GAE)	4.80-5.40
Antioxidant activity (mM Trolox equivalent TE)	0.29-0.35
Vitamin C (mg)	16-65

Table 3. Nutritional composition of *neera* (Misra 2016)

Parameters	Values
Total solids (g 100 ml^{-1})	15.2-19.7
pH	3.9-4.7
Specific gravity	1.058-1.077
Total sugars (g 100 ml^{-1})	14.40
Original reducing sugars (g 100 ml^{-1})	5.58
Total reducing sugars (g 100 ml^{-1})	9.85
Total ash (g 100 ml^{-1})	0.11-0.41
Citric acid (g 100 ml^{-1})	0.50
Iron (g 100 ml^{-1})	0.15
Phosphorus (g 100 ml^{-1})	7.59
Ascorbic acid (mg 100 ml^{-1})	16-30
Total protein (g 100 ml^{-1})	0.23-0.32
Alcohol (%)	Nil

A large amount of free amino groups, released due to the catabolism of amino acids like glutamine and asparagine, reacts with the carbonyl group of reducing sugars present in the sap, forming flavouring compounds through a series of secondary reactions (Hebbar et al., 2018). The distinctive taste of the sap arises from these flavor compounds. Borse et al., (2007) examined the volatile compounds in the sap and identified substances such as 1-hexanol, ethyl lactate, tetradecanone, phenyl ethyl alcohol, 2-methyl tetrahydrofuran (THF), and 3-hydroxy-2-pentanone.

The inflorescence sap (*neera)* contains high sugar content (10–15%) and, if not collected hygienically, it quickly ferments, turning into toddy. During this fermentation process, the sugar converts into alcohol, resulting in a beverage with an alcohol content of 5–8% (Iwuoha and Eke 1996). The level of reducing sugar serves as an indicator of the sap's freshness, and it should ideally be below 1.0 g100 ml^{-1}. *Neera* contains16 types of amino acids and a variety of vitamins, such as vitamin C and the B complex vitamins, with a particular abundance of nicotinic acid (Aalbersberg et al., 1997). The sap primarily contains vitamin C as its major vitamin, alongside the vitamin B complex (Hebbar et al.2015). The sap also contained various vitamins C, B1, B2, B3, B4, and B10 (Asghar et al., 2020b). Femi et al., (2020) also reported that *neera* contains vitamins (A and C), and has antioxidant properties thereby preventing damage or death of cells. The nutrient-rich components of coconut inflorescence sap make it highly prone to spontaneous fermentation, particularly during harvesting and when exposed to sunlight. *Neera* also has medicinal properties, and hence being used as a syrup in Ayurvedic system of medicine (Somashekaraiah et al., 2019).

Neera collected using the coco-sap chiller (*Kalparasa*) exhibited a higher pH and Brix percentage, almost no alcohol content, and lower bacterial and yeast populations compared to *neera* traditionally collected in earthen pots. In addition to its high sugar content, *Kalparasa* contains amino acids, total phenols, flavonoids, and antioxidants at levels that are 2.5, 1.5, 4.6, and 1.8 times higher, respectively, than those found in *neera*. Additionally, products derived from *Kalparasa*, such as sap concentrate and sugar, are rich in amino acids, polyphenols, flavonoids, vitamins, and antioxidants (Hebbar et al., 2015).

Barh and Mazumdar (2008) discovered that coconut sap was the most mineral-rich, containing 6.95 mg100 ml^{-1} of sodium (Na) and 3.16 mg 100 ml^{-1} of potassium (K), compared to sugar palm juice and date palm sap. Additionally, coconut sap showed the highest fermentation potential, producing 3.4% alcohol among the saps studied. Joelle et al., (2017) examined the mineral content of sap from two coconut varieties (MYD, WAT) and two hybrids (PB121+ and PB113+) grown in Côte d'Ivoire. The ash content ranged from 0.18% to 0.27% (w/w), with MYD (0.26%) and PB113+ (0.27%) having higher ash content than WAT and PB121+. The mineral composition included eight macro elements (K, Cl, Si, Na, Mg, P, S, and Ca) and five trace elements (Fe, Cu, Mn, Zn, and Br). Macro element levels varied between 1.25 mg100 g^{-1} (Ca in PB113+) and 90.65 mg100 g^{-1} (K), while trace elements ranged from traces to 0.70 mg100 g^{-1}. The sap from MYD and PB113+ showed excellent mineral composition, particularly high in K, P, Fe, Si, Na, S, and Mg, whereas WAT sap had lower mineral content.

Table 4. Mineral composition of coconut sap (Hebbar et al., 2018)

Sodium (mg)	69.40-117.50
Potassium (mg)	146.10-182.40
Calcium (mg)	1.62
Manganese (mg)	0.009-0.014
Magnesium (mg)	6.00
Copper (mg)	0.03-0.04
Zinc (mg)	0.02-0.03
Iron (mg)	0.05-0.06

Table 5. Mineral composition of coconut sap (Femi et al.2020)

Minerals	Value
Sodium (mg)	90.64 ± 1.32
Potassium (mg)	167.34 ± 2.74
Copper (mg)	0.031 ± 0.10
Manganese (mg)	0.016 ± 0.04
Phosphorus (mg)	3.8 ± 0. 63
Iron (mg)	0.056 ± 0.43
Zinc (mg)	0.020 ± 0.04

Analysing the chemical properties of coconut sap collected from Negeri Sembilan, Malaysia, Asghar et al., (2020) found predominance of minerals (mg l^{-1}) K (960.87), Na (183.21), and Mg (22.91), Fe (1.36), Ca (0.42), Zn (0.338), Mn (0.105), and Cu (0.065). The mineral composition of coconut sap as reported by various researchers is presented in Tables 4 and 5.

Despite the sap's high nutritional value and health benefits, there is a lack of awareness among people about consuming this drink (Chinnamma et al., 2019). Dodecanoic acid and palmitoleic acid, along with ethanol and ethyl esters, render the sap unpalatable after fermentation. Conversely, ethyl lactate, 3-hydroxy-2-pentanone, and farnesol are typically found in both fresh and fermented sap.

The mineral nutrition of inflorescence sap of four coconut varieties (MYD, Keraganga, WCT, and Kerasree) was evaluated (Raghu and Joseph 2020). The sap from WCT record the highest values for K (1.4%) and S (0.3%), while MYD registered the highest values for P (0.43%), Ca (579.20 mg l^{-1}), and Cu (2.31 mg l^{-1}). Sap from WCT also registered higher values (mg l^{-1}) for Mg (30.76), Fe (96.68), Mn (1.43), as well as Zn (7.13). Protein-derived nitrogen in the sap did not differ among the varieties, whereas, the non-protein nitrogen was absent in the sap from any of the varieties.

5. Value-Added Products of Coconut Palm Sap

5.1. Coconut Sugar

Indonesia and the Philippines are the leading producers of palm sugar globally, which is made by caramelizing the sap (Saputro et al., 2019). It is a key ingredient in Indonesian sweet soy sauce and intermediate moisture meat products, contributing a distinctive taste and flavour to these items (Purnomo

2007). The unique flavour profile of coconut sugar has also made it a popular choice as a flavouring agent in confectionery and baked goods. It is brown, caramel aroma and contains 2-3% moisture, plant-based, natural sweetener that undergoes minimal processing. It is an excellent choice for many vegan diets, offering a healthier alternative (Figure 1) (Hebbar et al., 2020). Exports of products made from palm sugar or coconut sap reached 36.5 thousand tons, with a value of US$49.3 million in 2019. These export products must adhere to the food regulations of the destination countries, such as those set by the European Food Safety Authority (EFSA) for European countries or the Food and Drug Administration (FDA) for the United States. Mixing palm sugar with corn sugar has been shown to slow digestion, resulting in a lower GI compared to refined cane sugar. Additionally, palm sugar contains amino acids, vitamins (B1, B2, B3, and B6), antioxidants (2,3-dihydro-3,5-dihydroxy-6-methyl-4(H)-pyran-4-one), and dietary fiber (inulin), all of which contribute to various health benefits (Kurniawan and Kustiningsih 2019).

Figure 1. Coconut sugar derived from inflorescence sap utilizing double-jacketed cooker.

5.2. Coconut Jaggery

Coconut jaggery is made by concentrating unfermented coconut sap, which is commonly referred to as *'Gur,' 'Karupatti,' 'Vellam,' 'ThaatiBella,'* and *'Olebella'* in India. It serves as a low-calorie natural sweetener and aids in digestion. The sap, which contains approximately 80% water is evaporated to remove the water content. Before heating, the sap is filtered to eliminate impurities. Upon evaporation, a thick substance is obtained, which, with further heating, undergoes crystallization. Upon cooling, it solidifies into a dense form. The resulting product exhibits a dark colour due to sugar caramelization (Mithra et al., 2013). This translucent brown semi-solid structure, obtained through caramelization, has a pH ranging between 6.5 and 7.0 (Nighitha and Sudagar 2017). The relatively high cost of palm jaggery is due to its medicinal properties and its content of minerals such as calcium, phosphorus, and iron.

Coconut Sap (pH – 7)
(100 litres)

↓

Straining ------------ **Foreign matter rejected**

↓

Boiling ------------ **Removing the scum**

↓

Concentration ---------- **Water removed**

↓

Pour hot into moulds
prepared out of coconut leaves

↓

Cooling at room temperature

↓

Jaggery (15 kg)

Figure 2. Flow diagram depicting the preparation of coconut jaggery.

When coconut sap is transformed into a solid or semi-solid crystalline mass for direct consumption, it is called palm jaggery, which has a distinct aroma. Traditionally, palm jaggery is made by boiling raw sap in large, shallow, round-bottomed vessels. The process begins by heating the fresh sap to 40°C in a pan and neutralizing the pH by slowly adding phosphoric acid or a solution of triple superphosphate while stirring continuously. Afterward, the juice is filtered and boiled in open-fired furnaces using metal pans of 20 to 24 gauge thickness. As the sap boils, froth and foam rise to the surface and are skimmed off using a perforated ladle. The juice is occasionally stirred to ensure proper mixing and promote rapid evaporation. Once the juice thickens, the heat is carefully controlled to avoid caramelization. To check if the mixture has reached the correct consistency, a small amount is placed in water and rolled into a ball. If the ball forms into a hard shape, the mass is ready to be molded. Thermometers can also be used to monitor the temperature. Before placing the thickened sap into molds, the molds can be moistened with water or coated with fresh sweet oil to make removal easier. After the jaggery has set, it is removed from the molds and packed. High-quality jaggery is typically hard, crystalline, and golden in color. The following (Figure 2) is a flow diagram of the coconut jaggery preparation process.

Liquid palm jaggery can be prepared in liquid form or semi solid form. It is very fine and chemical free liquid syrup.

5.3. Coconut Syrup/Honey

Coconut sap honey (Figure 3), a value-added product, derived from the sap of coconut inflorescence, has a pH between 6.5 and 7. It is made by boiling the sap over low to moderate heat until the temperature reaches 110°C, resulting in a sticky consistency with 80-85°Bx (Brix). Care must be taken to avoid prolonged heating, which can lead to charring and affect the honey's color. Once the desired sticky texture is achieved, the coco honey or syrup is cooled and poured into appropriate containers, and stored in a refrigerator. Coconut sap honey can maintain its quality for up to one year under refrigerated conditions. It is commonly used as table syrup for sweetening confectionery items like ice cream and is a rich source of iron, making it valuable for anemic patients and also in pharmaceutical formulations.

Figure 3. Coconut Honey derived from coconut inflorescence sap.

A flow diagram outlining the preparation process of coconut sap honey is given below (Figure 4).

Coconut Sap (1 litre)

↓

Filtration ------------ Foreign matter rejected

↓

Boiling ------------ Removing the scum

↓

Concentration to honey consistency ------- Water removed

↓

Coconut Honey (100 ml) -------- 75 – 80 ⁰ Brix

↓

Cooling and bottling

Figure 4. Flow diagram showing the preparation of coconut honey,

Srikaeo and Thongta (2015) produced coconut syrup by heating it continuously at temperatures ranging from 97°C to 120°C for approximately 3.5 h, occasionally stirring throughout the heating process. Heating was halted once viscous and fairly thick hot syrup with 75°-80° Brix was attained. The resulting product was then cooled to produce coconut sugar syrup (honey).

It is abundant in essential minerals like potassium, magnesium, zinc, and iron, and serves as a natural source of essential vitamin B complex and vitamin C. Its primary sugars are fructose (38-40%) and glucose (33-35%), contributing to its low GI and suitability for individuals with diabetes. Furthermore, *neera* honey's high sugar content and low glass transition temperatures make it notably sticky (Pandiselvam et al., 2023).

5.4. Vinegar

Vinegar is primarily produced through the fermentation of rice, vegetables, malt, fruits, and tea. Palm vinegar, on the other hand, has multiple utilities including medicinal applications, acting as a food additive and preservative, as well as in various culinary uses. The initial stage of vinegar production involves alcohol fermentation, succeeded by acetic acid fermentation. Palm vinegar is suggested to possess superior export potential compared to synthetic vinegar (Samuel et al., 2016). The flavour and aroma of vinegar are attributed to the acetic acid generated during fermentation. Studies on 8-10 weeks old male Sprague-Dawley rats demonstrated that the aqueous extract of palm vinegar exhibited hypoglycemic effect (Yusoff et al., 2015).

5.5. Toddy

The indigenous microorganisms found in the sap initiate fermentation, transforming the sweet sap into a milky white liquid known as palm wine, with an alcohol content ranging from 4 to 4.5% (Lasekan and Abbas 2010). This traditional alcoholic beverage derived from the Palmae family is called "toddy" and is popular in tropical regions across Asia, Africa, and South America (Figure 5). Due to its alcohol content, the product is viewed as a potential substitute for disinfectants. However, its high sugar concentration renders it unsuitable for use in clinical or household settings. Studies have indicated its antibacterial effects against *Staphylococcus aureus* and *Pseudomonas aeruginosa*.

Figure 5. Fermented coconut inflorescnece sap (toddy).

5.6. Levan Production

Palm sap is recognized as a favourable substrate for microbial levan polysaccharide production. Researchers explored the potential of fresh coconut inflorescence sap as a fermentation medium for levan production using *Bacillus subtilis*. The yield of levan is significantly influenced by fermentation conditions such as temperature, pH, and agitation speed. Therefore, palm sap presents a cost-effective source for levan polysaccharide production and offers faster production times. The organic nature of palm sap provides natural biochemicals that enhance microbial growth, making it an ideal fermentation medium without the need for additional chemical additives (Mummaleti et al., 2022).

5.7. Other Value-Added Products

A variety of enhanced products such as candies, chocolates, cookies, soft drinks, squash, etc. can be produced from coconut sap. Confectionery items, such as cookies crafted from the sap, can serve as suitable snacks for individuals with diabetes when supplemented with ingredients like multigrain, oats, whole wheat, etc. Kalpa bar, a chocolate infused with coconut sugar, is crafted exclusively from plant-based ingredients, completely devoid of dairy components. The formulation of dark chocolate with coconut sugar was fine-tuned to include 45% cocoa liquor, 30% coconut sugar, and 25% cocoa butter, resulting in a product rich in polyphenols, flavonoids, antioxidants, and minerals (Hebbar et al., 2022). Kalpa drinking chocolate, another delightful offering, is prepared using low GI coconut sugar and premium cocoa powder. In West Bengal sweets, coconut sugar is used in place of other sweeteners, utilizing the concentrated form of coconut inflorescence sap. These sweets provide essential minerals, vitamins, and beneficial fibers, with a low GI suitable for both health-conscious consumers and those seeking a sweet indulgence (Ghosh et al., 2018). Additionally, Frozen Coconut Delicacy, made from coconut milk, coconut sugar, tender coconut water, and pulp, benefits from the addition of coconut sugar, enhancing its flavour and overall appeal (Beegum et al., 2022).

6. Preservation of Palm Sap

Many preservation techniques have been developed to prolong the shelf life of palm sap. The collected sap undergoes spontaneous fermentation caused by microorganisms in the surrounding environment of the open container. The self-fermentation transforms palm sap into an alcoholic beverage known as toddy (Hebbar et al., 2018). Researchers have tried to preserve the sap using various physical and chemical processing techniques. Lime is a common method of inhibiting the fermentation of sap by coating the interior of the collection container. Despite the lime coating, the sap undergoes partial fermentation, making it unsuitable for marketing as a health beverage (Hebbar et al., 2015). ICAR-CPCRI has developed the 'Coco sap chiller' to collect unfermented *neera*, and the sap so collected is fresh, hygienic, and non-alcoholic. *Kalparasa* is the trade name for this unfermented coconut sap.

Thermal methods like pasteurization and sterilization are commonly used to prolong the shelf life of palm sap. The pasteurization of palm sap in India is typically carried out between 3 to 7 min. at 90 to 95°C. In Thailand, pasteurization was typically completed at 60 to 70°C for 60 minutes (Naknean and Meenune 2015). The impact of microwave treatment (600 and 900 W) on the physicochemical properties of coconut inflorescence sap under refrigerated conditions was investigated by Pandiselvam et al., (2020). It was found that using 900 W for 120 sec preserved the physicochemical properties for up to 16 days. Hebbar et al., (2018) found that palm sap preservation can be achieved by subjecting it to pasteurization at temperatures ranging from 90 to 95°C for a duration of 3 to 7 min. or alternatively by boiling it at 60 or 70°C for 60 min. Coconut sap preservation was accomplished through the utilization of different fermentation inhibitors derived from sources, namely *Vateria copallifera*, *Careya arborea*, *Azadirachta indica*, and lime.

7. Health Benefits

As palm trees are abundant in nutrients, the sap from their inflorescences provides many health benefits. The reduced GI of palm sap is associated with controlled glucose absorption into the bloodstream, aiding in the management of diabetes mellitus and cancer by reducing the requirement for insulin. The active constituents present in palm sap have demonstrated a protective effect on the pancreas against alloxan-induced cytotoxicity. This protection is achieved by enhancing glucose uptake, accelerating liver glycolysis, and restoring the population of beta cells (Sarma et al., 2022). Vitamins and electrolytes found in the sap serve as co-enzymes and co-factors in myriad metabolic processes that protect the body from various diseases and improve immunity (Jose et al., 2017). Phytochemicals and antioxidants present in palm sap aid in preventing the formation of various cancers by protecting cells against damage. Due to its abundant antibacterial and antioxidant properties, the sap is beneficial for shielding the skin against infections and oxidative stress induced by free radicals. Palm sap contains significant amounts of thiamine and inositol, essential nutrients for the proper functioning of the nervous system. Phosphorus in sap facilitates the body's absorption of calcium (Asha et al., 2019). Because palm sap is classified as a low GI food, it aids in decreasing the fat levels in the bloodstream, elevating high-density lipoprotein (HDL) levels, and subsequently reducing the risk of blood vessel clot formation. Moreover, the presence of ascorbic acid in the sap mitigates the oxidation of low-density lipoproteins (LDL) and inhibits the formation of free radicals, thereby lowering the occurrence of cardiovascular diseases (Ratheesh et al., 2017).

Conclusion

Coconut palm sap can be relished as a refreshing non-alcoholic beverage with a great deal of health benefits. The extraction of sap directly from the tree poses a significant challenge to the industry. Due to its high microflora content, the sap is susceptible to rapid fermentation immediately following harvest. The development of coco-sap chiller technology at ICAR-Central Plantation Crops Research Institute (CPCRI) has largely solved the problem of collecting hygienic, unfermented sap from palm trees. The collected sap, being highly nutritious and healthy, is suitable for direct sale as fresh juice when kept refrigerated. Alternatively, it can be processed into products like coconut sugar, jaggery, nectar, or syrup without adding chemicals, preservatives, or additives. A major hurdle for the industry is the cost of processing the sap on a commercial scale. In light of the coconut sap's benefits and challenges, future research should focus on its commercialization.

References

Aalbersberg, B., Singh, R.R.S., Praveen Ravi, P.R. (1997). Nutrient analysis of coconut toddy. *Trop Sci* 37:160-163.

Asghar, M.T., Yusof, Y.A., Noriznan, M., Yaacob, M.E., Ghazali, H.M., Chang, L.S., Manaf, Y.N. (2020b) Effect of processing method on vitamin profile, antioxidant properties and total phenolic content of coconut (*Cocos nucifera* L.) sugar syrup. Int J *Food Sci Technol* 55(7), doi: 10.1111/ijfs.14529.

Asghar, M.T., Yusof, Y.A., Mokhtar, M.N., Ya'acob, M.E., Ghazali, H.M., Chang, L.S., Manaf, Y.N. (2020a) Coconut (*Cocos nucifera* L.) sap as a potential source of sugar: antioxidant and nutritional properties. *Food Sci Nutr* 8:1777–1787, https://doi.org/10.1002/fsn3.1191.

Asha, S., Ratheesh, M., Jose, S.P., Krishnakumar, I.M., Sandya, S. (2019). Neera : A non-alcoholic nutritious beverage from unopened inflorescence of coconut palm. In: Natural beverages. Elsevier, pp 339–360.

Azlan, A., Khoo, H.E., Sajak, A.A.B., Kadir, N.A., Yusof, B.N.M., Mahmood, Z., Sultana, S. (2020). Antioxidant activity, nutritional and physicochemical characteristics, and toxicity of minimally refined brown sugar and other sugars. *Food Sci Nutrn* 8(9): 5048–5062. https://doi.org/10.1002/fsn3.1803.

Barh, D., Mazumdar, B. (2008). Comparative nutritive values of palm saps before and after their partial fermentation and effective use of wild date (*Phoenix sylvestris* Roxb.) sap in treatment of anemia. *Res J Med Med Sci* 3(2):173-176.

Beegum, P.P.S., Nair, J.P., Manikantan, M.R., Pandiselvam, R., Shil, S., Neenu, S., Hebbar, K.B. (2022). Effect of coconut milk, tender coconut and coconut sugar on the physicochemical and sensory attributes in ice cream. *Food Sci Technol* https://doi.org/10.1007/s13197-021-05279-y.

Borse, B.B, Rao, L.J.M., Ramalakshmi, K., Raghavan, B. (2007). Chemical composition of volatiles from coconut sap (neera) and effect of processing. *Food Chem* 101 (3): 877–880.

Chinnamma M., Bhasker S., Hari B.M., Sreekumar D., Madhav H. (2019). Coconut neera - a vital health beverage from coconut palms: harvesting, processing and quality analysis. *Beverages* 5 (1):22, 10.3390/beverages5010022.

Divya, N.M., Nandhini, G., Nabish, S., Srinithi, S., Kalidas, K. (2019). Neera and its importance - A review. Int J Recent *Sci Res* 10 (2): 30302–30306.

Femi, B.M.R., Malhija, N., Sherin Mary, T., Kumaran, T. (2020) Coconut (*Cocos nucifera)* inflorescence sap nutritional characteristics and health effects. *J Basic Appl Res Biomed* 6(1): 50-52.

Ghosh, D. K., Bandyopadhyay, A., Das, S., Hebbar, K.B., Biswas, B. (2018). Coconut sap (Neera)-untapped opportunity of spinoff gains in West Bengal, India. *Int J Current Microbiol Applied Sci* 7 (9): 1883–1897.

Gopal, M., Shil, S., Gupta, A., Hebbar, K.B., Arivalagan, M. (2021). Metagenomic investigation uncovers presence of probiotic-type microbiome in Kalparasa®(fresh unfermented coconut inflorescence sap). *Frontiers Microbiol* 12, 662783.

Hebbar, K.B., Ramesh, S.V., Ghosh, D.K., Beegum, P.P.S., Pandiselvam R., Manikantan M.R., Mathew A.C. (2022). Coconut sugar-a potential storehouse of nutritive metabolites, novel bio-products and prospects. *Sugar Tech* 24 (3): 841–856.

Hebbar, K.B., Arivalagan, M., Manikantan, M.R., Mathew, A.C., Chowdappa, P. (2015). Kalparasa collection and value addition. Technical Bulletin No.92, ICAR-CPCRI, Kasaragod, Kerala, 28p.

Hebbar, K.B., Arivalagan, M., Manikantan, M.R., Mathew, A.C., Thamban, C., Thomas G.V., Chowdappa P. (2015). Coconut inflorescence sap and its value addition as sugar–collection techniques, yield, properties and market perspective. *Current Sci* 108(9):1411–1417.

Hebbar, K.B., Arivalagan, M., Pavithra, K.C., Roy, T.K., Gopal, M., Shivashankara, K.S., Chowdappa, P. (2020). Nutritional profiling of coconut (*Cocos nucifera* L.) inflorescence sap collected using novel coco-sap chiller method and its value added products. *J Food Measurement and Characterization* 14:2703–2712.

Hebbar, K.B., Pandiselvam, R., Manikantan, M.R., Arivalagan, M., Beegum, P.P.S., Chowdappa, P. (2018). Palm sap - quality profiles, fermentation chemistry, and preservation methods. *Sugar Tech* 20 (6):621–634.

Hori, K., Surjoseputro, S., Purnomo, H., Foe, K., Hashimura, F. (2001). Indigenous technology of coconut sugar production in the villages of Genteng Banyuwangi (East Java) and Dawan, Klungkung (Bali) and the knowledge about palm sugar by Japanese young people. Bulletin of Fukuoka University of Edn 50:109- 118.

Iwuoha, C.I., Eke, O.S. (1996). Nigerian indigenous fermented foods: their traditional process operation, inherent problems, improvements and current status. *Food Res Int* 29 (5–6): 527–540.

Joelle, O.D.M., Ysidor, K.G., Louis, K.K.J., Rachel, A.R., Marius, B.G.H. (2017) Mineral nutrients in inflorescence sap produced from various coconuts (*Cocos nucifera* L.) cultivated in Côte d'Ivoire. *J Exp Agric Int*16(2): 1-9, doi: 10.9734/JEAI/2017/32557.

Jose, S.P., Asha, S., Krishnakumar, I.M., Ratheesh, M., Santhosh, S., Sandya, S., Kumar, G., Pramod, C. (2017). Nephro-protective effect of a novel formulation of unopened coconut inflorescence sap powder on gentamicin induced renal damage by modulating oxidative stress and inflammatory markers. *Biomed Pharmacother* 85:128–135.

Kurniawan, T., Kustiningsih, I. (2019). Phenolic analysis and characterization of palm sugar (*Arenga pinnata*) produced by the spray dryer. *Oriental J Chem* 35 (1): 150.

Lasekan, O., Abbas, K.A. (2010). Flavour chemistry of palm toddy and palm juice: a review. *Trends Food Sci Technol* 21 (10): 494–501.

Leelayuwat, N., Saelim, B., Kanpettha, Y., Aneknan, P.T., Tong-U.N. T. (2018). Effects of organic coconut flower syrup on glucose and insulin responses and substrate utilization. *Asia Pac J Sci Technol* 23 (4): 1–6, https://doi.org/10.14456/apst.2018.16.

Misra, B. (2016). Neera: The coconut sap: A review. *Int J Food Sci Nutrn* 1(4):35–38

Mithra, A., Swamy G. J., Chandrasekar V., Shanmgam S. (2013). Coconut value-added products. *J Indian Food Industry Mag* 32 (6): 29–36.

Mummaleti, G., Sarma, C., Kalakandan, S.K., Gazula, H., Sivanandham, V., Anandharaj A. (2022). Characterization of levan produced from coconut inflorescence sap using *Bacillus subtilis* and its application as a sweetener. *LWT*, 154, 112697.

Nakamura, S.I., Watanabe, A., Chongpraditnum, P., Suzui, N., Hayashi, H., Hattori, H., Chino, M. (2004) Analysis of phloem exudate collected from fruit-bearing stems of coconut palm: Palm trees as a source of molecules circulating in sieve tubes. *Soil Sci Plant Nutrn* 50(5):739-745. https://doi.org/10.1080/00380768.2004.10408530.

Naknean, P., Juntorn, N., Yimyuan, T. (2014). Influence of clarifying agents on the quality of pasteurised palmyra palm sap (*Borassus flabellifer* L.). *Int J Food Sci Technol* 49 (4): 1175–1183.

Naknean, P., Meenune, M. (2015). Impact of clarification of palm sap and processing method on the quality of palm sugar syrup (*Borassus flabellifer* Linn.). *Sugar Tech*, 17: 195–203.

Naknean, P., Meenune, M., Roudaut, G. (2010). Characterization of palm sap harvested in Songkhla province, Southern Thailand. *Int Food Res J*.17(4):977–986.

Nathanael, W.R.N. (1966). A tapping technique for the coconut palm. *J Agric Food Sci* 4: 87-99.

Nguyen, V.D., Harifara, R., Shiro, S. (2016). Sap from various palms as a renewable energy source for bioethanol production. *Chemical Industry and Chemical Eng Quart* 22 (4): 355–373.

Nighitha, M.T., Sudagar, I.P. (2017). Engineering properties of palmyrah palm jaggery (*Borassus flabellifer* L.). *Trends in Biosci* 10 (43):8974–8976.

Nurhadi, B., Sukri, N., Sugandi, W.K., (2018) Comparison of crystallized coconut sugar produced by traditional method and amorphous coconut sugar formed by two drying methods: vacuum drying and spray drying. *Int J Food Properties* 21(1):2339–2354, doi: 10.1080/10942912.2018.1517781.

Ouchemoukh, S., Louaileche, H., Schweizer, P. (2007). Physico-chemical characteristics and pollen spectrum of some Algerian honey. *Food Control* 18:52-58.

Pandiselvam, R., Hebbar, K.B., Manikantan, M.R., Prashanth, B.K., Beegum, S.P.P., Ramesh, S.V. (2020). Microwave treatment of coconut inflorescence sap (Kalparasa®): a panacea to preserve quality attributes. *Sugar Tech* 22:718–726.

Pandiselvam, R., Joseph, L.T., Manikantan, M.R., Khanashyam, A.C., Beegum, P.P.S., Ramesh, S.V. Balasubramanian, D., Neenu, S., Gopal, M., Mathew, A.C. (2023). Physical, chemical and functional attributes of Neera honey infused extrudates. *Bioeng* 10 (1): 114.

Priya, V.V., Balakrishnan, M., Karthikeyan, S., Rajkumar, P., Manikantan, M.R., Gurusamy, K., Pandiselvam, R. (2023). Evaluation of effect of tapping frequency and lime coating on the biochemical changes and fermentation rate of coconut inflorescence sap. *Sugar Tech* 25:906-915, https://doi.org/10.1007/s12355-023-01254-3.

Purnomo, H. (1992) Sugar components of coconut sugar in Indonesia. *ASEAN Food J* 7(4): 200–201.

Purnomo, H. (2007). Volatile components of coconut fresh sap, sap syrup and coconut sugar. *ASEAN Food J* 14 (1): 45.

Raghu, R.S., Joseph, B. (2020) Variability among the coconut varieties with respect to nutritional qualities and total mineral content in inflorescence sap. *Int J Chem Studies* 8(2): 2231-2234, https://doi.org/10.22271/chemi.2020.v8.i2ah.9080.

Ratheesh, M., Svenia, J.P., Asha, S., Sandya, S., Girishkumar, B., Krishnakumar, I.M. (2017). Anti-inflammatory effect of a novel formulation of coconut inflorescence sap against ox-LDL induced inflammatory responses in human peripheral blood mononuclear cells by modulating TLR-NF-κB signaling pathway. *Toxicology Mechanisms and Methods* 27 (8):615–621.

Redhead, J.F. (1989). Utilization of Tropical Foods: Tropical Oil-Seeds. *In FAO Food and Nutrition* 43 (3): 52.

Samuel, O., Lina, J., Ifeanyi, O. (2016). Production of vinegar from oil-palm wine using *Acetobacter aceti* isolated from rotten banana fruits. *Universal J Biomed Eng* 4 (1):1–5.

Saputro, A.D., Van de Walle, D., Dewettinck, K. (2019). Palm sap sugar: a review. *Sugar Tech* 21: 862–867.

Sarma, C., Mummaleti, G., Sivanandham, V., Kalakandan, S., Rawson, A., Anandharaj, A. (2022). Anthology of palm sap: The global status, nutritional composition, health benefits & value added products. *Trends Food Sci Technol* 119:530–549.

Sharma, A., Meghwal, M. (2021). An update on uses, benefits and potential application of Neera. *J. Nutrn Food Processing* 4(8): doi:10.31579/2637-8914/072, 18 pages.

Somashekaraiah, R., Shruthi, B., Deepthi, B.V., Sreenivasa, M.Y. (2019). Probiotic properties of lactic acid bacteria isolated from neera: a naturally fermenting coconut palm nectar. *Frontiers Microbiol* 10:1382, https://doi.org/10.3389/fmicb.2019.01382.

Srikaeo, K., Thongta, R. (2015) Effects of sugarcane, palm sugar, coconut sugar and sorbitol on starch digestibility and physicochemical properties of wheat based foods. *Int Food Res J* 22(3): 923–929.

Sudha, R., Niral, V., Hebbar, K.B., Samsudeen, K. (2019). Coconut inflorescence sap. *Current Sci* , 116 (11):1809–1817.

Tomomatsu, A., Itoh, T., Wijaya, C.H., Nasution, Z., Kumendong, J., Matsuyama, A. (1996). Chemical constituents of sugar-containing sap and brown sugar from palm in Indonesia. *Japanese J Tropical Agric* 40 (4):175–181.

Trinidad, T.P., Mallillin, A., Avena, E., Rodriguez, R., Borlagdan, M., Cid, K.B., Biona, K. (2015). Coconut sap sugar and syrup: a promising functional food/ingredient. *Acta Manilana* 63: 25–32.

Trinidad, T.P., Mallillin, A.C., Sagum, R.S., Encabo, R.R. (2010). Glycemic index of commonly consumed carbohydrate foods in the Philippines. *J Functional Foods* 2 (4):271–274.

Xia, Q., Li, R., Zhao, S., Chen, W., Chen, H., Xin, B., Huang, Y., Tang, M. (201)1. Chemical composition changes of post-harvest coconut inflorescence sap during natural fermentation. *African J Biotechnol* 10 (66):14999–15005.

Ysidor, K.N., Rachel, A.R., Louis, J.K.K., Muriel, O.D.J., Prades, A., Kouassi, A., Marius, B.G.H. (2014) Glucide factors of the inflorescence sap of four coconut (*Cocos nucifera* l.) cultivars from Côte D'ivoire. *Int J Biochem Res Review* 4(2): 116-127.

Yusoff, N.A., Yam, M.F., Beh, H.K., Razak, K.N.A., Widyawati, T., Mahmud, R., Ahmad, M., Asmawi, M. Z. (2015). Antidiabetic and antioxidant activities of *Nypa fruticans* Wurmb. vinegar sample from Malaysia. *Asian Pacific J Trop Med* 8 (8): 595–605.

Zafar, M.A., Mills, K.R., Zheng, J., Regmi, A., Hu, S., Gou, L., Chen, L. (2019) Low-glycemic index diets as an intervention for diabetes: a systematic review and metaanalysis, *Am J Clin Nutr* 110:891–902. https://doi.org/10.1093/ajcn/nqz149.

Chapter 5

Advances in Coconut Palm Sap Collection

R. Pandiselvam*
M. R. Manikantan
S. V. Ramesh
and K. B. Hebbar
Division of Physiology, Biochemistry and Post Harvest Technology,
ICAR-Central Plantation Crops Research Institute (ICAR-CPCRI),
Kasaragod, Kerala, India

Abstract

The collection of coconut palm sap has traditionally relied on manual methods, which are labour-intensive and often inefficient. Recent advancements in technologies such as robotics, machine vision, artificial intelligence (AI), and the Internet of Things (IoT) offer significant opportunities to revolutionize this process. This chapter explores the current practices and emerging technologies for coconut palm sap collection, focusing on the integration of robotic systems and AgroBots, as well as the application of machine vision and AI for enhancing collection efficiency and monitoring sap quality. The role of IoT and Big Data in monitoring and optimizing sap collection processes is also discussed, highlighting their potential to provide real-time data and predictive analytics for improved management. This chapter addresses the prospects of these technologies, including the potential for reduced labour costs, increased sap yield, and enhanced quality control. The challenges associated with high implementation costs, the need for reliable power sources and connectivity in remote areas, and the

---

* Corresponding Author's Email: r.pandiselvam@icar.gov.in

In: Coconut Palm Sap
Editors: K. B. Hebbar, V. Krishnakumar and S. V. Ramesh
ISBN: 979-8-89530-640-6

complexity of data management are also considered. By merging insights from recent research and technological advancements, this chapter provides a comprehensive overview of the potential and challenges in modernizing coconut palm sap collection, offering valuable perspectives for future developments in the field.

Keywords: AgroBots, machine vision, artificial intelligence, big data, sap collection, predictive analytics

1. Introduction

Coconut palm sap, also known as *neera*, is a sweet, translucent liquid obtained by tapping the unopened spadix of coconut palms. The traditional method of sap collection involves the manual cutting of the spadix and the attachment of a container to collect the sap (Dalibard 1999). This labour-intensive process requires skilled workers to climb the palms daily (twice a day), often resulting in variations in yield due to factors such as environmental conditions, the skill of the tapper, and the time of day the sap is collected (Hai et al., 2024; Hebbar et al., 2015; Hebbar et al., 2018). The sap is highly perishable, necessitating prompt collection and processing to prevent fermentation, which can lead to spoilage and a decline in quality (Ellsäßer et al., 2020; Ezzatzadegan et al., 2021; Pandiselvam et al., 2021).

Despite the challenges, the method remains widely used to collect the sap is insulated box or "coco-sap chiller" technology developed by ICAR-CPCRI, particularly in regions where coconut palms are abundant (Hebbar et al., 2018). However, the demand for increased efficiency, consistency in yield, and the need for non-fermented sap have driven research into more advanced and simple methods of sap collection (Ellsäßer et al., 2020; Ranasinghe and Waidyanatha 2003).

This chapter discusses the advancements made in coconut palm sap collection and processing by integrating modern technologies. It aims to review traditional and current practices, explore the application of robotics, machine vision, and artificial intelligence to enhance collection efficiency, and evaluate the role of IoT and Big Data in optimizing sap quality and predicting yield. The chapter also identifies key challenges and opportunities associated with these technologies and propose future directions for research and development to further improve sap collection and processing.

2. Robotics-Based Collection and AgroBots

The integration of robotics into agriculture has been gaining popularity, with significant advancements in the development of robotic systems designed to perform tasks that were traditionally performed manually. In the context of coconut palm sap collection, robotic systems can offer a solution to the challenges posed by the labour-intensive nature of traditional tapping methods (Batz et al., 2023). These systems can be designed to perform the tapping process with precision, consistency, and minimal human intervention, potentially leading to higher yields and better quality sap (Alander et al., 2013; Okere et al., 2021).

Robotic systems equipped with sensors can monitor the sap flow and adjust the tapping process in real-time, ensuring optimal sap collection. The use of AgroBots, or agricultural robots can further enhance this process by automating tasks such as the placement and retrieval of collection containers, sap transportation, and even the monitoring of palm health (Batz et al., 2023; Fahey et al., 2020; Rashid et al., 2021). This technology not only reduces the dependency on skilled labour but also minimizes the risks associated with manual sap collection, such as accidents and inconsistent yields (Alander et al., 2013).

However, the development and deployment of such systems encounter several challenges. The complexity of designing a robotic system capable of navigating the varying terrains and heights of coconut palms, complex arrangement of coconut leaf, unique shape and hardness of coconut inflorescence, coupled with the need for reliable power sources in remote locations present significant technical hurdles (Batz et al., 2023). Additionally, the cost of these systems may be unaffordable for small-scale farmers, limiting their extensive adoption (Alander et al., 2013).

3. Machine Vision Systems for Sap Collection and Quality Evaluation

Machine vision systems, which use cameras and sensors to capture and analyze visual information, are increasingly being applied in agricultural settings to improve the efficiency and accuracy of various processes (Alander et al., 2013). In the case of coconut palm sap collection, machine vision can be employed to monitor the sap flow, detect impurities, and assess the quality of the sap in real-time. These systems can be integrated into robotic collection systems to enhance their precision and effectiveness (Beltran Ortega et al., 2016; Fountas et al., 2020).

In machine vision systems for coconut palm sap collection, careful consideration of image processing parameters is essential for accurate monitoring. The camera should be positioned at a 45-degree angle relative to the collection point to effectively capture the sap droplets while minimizing environmental effects such as glare and shadows. Consistent, proper lighting is crucial to prevent reflections and ensure clear visibility of the sap, with LED lights offering adjustable intensity for uniform illumination. A dark background enhances the contrast between the sap and its surroundings, facilitating clearer detection of sap droplets. High-resolution cameras and a focus on sharp point, well-defined images are necessary to analyze droplet size and colour accurately. Advanced image processing algorithms, including colour threshold and pattern recognition, may be used to assess the sap's quality by detecting colour difference and impurities.

By detecting any deviations in the colour of the sap from the defined quality parameters, the image processing systems can trigger automated responses, such as adjusting the tapping process or alerting the operator to indicate potential issues (Beltran Ortega et al., 2016). This not only ensures that the sap collected is of the highest quality but also reduces the possibility of spoilage (Beltran Ortega et al., 2016).

Moreover, machine vision can play a crucial role in the preservation of sap. As the sap is highly susceptible to fermentation, real-time monitoring of its properties during collection and storage can help in taking immediate corrective actions to prevent quality degradation (Beltran Ortega et al., 2016). Capturing the drop-by-drop nature of sap flow presents unique challenges compared to continuous flow monitoring. The slow and irregular nature of coconut sap droplets requires precise synchronization of camera frames and image processing algorithms to accurately capture and analyze each droplet. Ensuring that the system is capable of handling this complexity is vital for successful implementation of sap collection processes.

Implementing machine vision systems for coconut sap collection in Kerala, India where the state experiences over 2500 mm of annual rainfall, presents challenges such as reduced visibility due to rain, poor lighting conditions, and reflections from water. The rainy season can obscure sap droplets, complicate image processing, and introduce background noise, and affecting the system accuracy. Additionally, high humidity can damage equipment, and inconsistent sap flow due to weather variations requires flexible technology. These factors must be carefully addressed to ensure effective and reliable operation during the rainy season.

In line with advancements in producing coconut inflorescence by-products and derivatives, characterizing their texture through spectral analysis has become a priority. Accurately determining the texture of coconut inflorescence is crucial for identifying its maturity, which plays a key role in harvesting larger quantities of high-quality sap. A study by Suganob et al., (2023) utilized MATLAB and developed five regression-based adaptive neuro-fuzzy inference system (ANFIS) models to predict five texture phenotypes of coconut inflorescence through spectral feature analysis. This study concluded that the five ANFIS models accurately predicted the corresponding textures, with the model for contrast prediction being the most accurate and the model for entropy texture prediction being the least accurate. However, further improvements are recommended by ensuring a more balanced number of images across categories in the dataset.

The quality of sap, which serves as the raw material, is a crucial factor in determining the quality of coconut sugar. Therefore, early detection of coconut spadices is essential for enhancing the yield of harvested sap for coconut sugar production. A study utilized the concepts of machine vision and deep learning through object detection for the effective identification of coconut inflorescences. The You Only Look Once v4-Cross Stage Partial connection (YOLOv4-CSP) algorithm, utilizing the Darknet53 convolutional neural network model, achieved the highest average precision value of 0.84, outperforming other object detection algorithms. The YOLOv4-CSP Darknet53 model effectively detected coconut inflorescences in images despite environmental challenges like variations in illumination and occlusion. The findings of this study can be applied to improve and optimize the harvesting of fresh coconut sap by enabling early detection of coconut inflorescences (Suganbob et al., 2022).

4. Prospects of Artificial Intelligence (AI) and Machine Learning

Artificial Intelligence (AI) and Machine Learning (ML) are rapidly transforming agriculture by enabling more precise and data-driven decision-making (Batz et al., 2023; Natarajan and Ponnusamy 2022). The process of collecting sap from palms is intricate, requiring tasks such as climbing the tree to slice the spadix, removing the sap from the container, replacing it with a fresh one, adding ice or gel packs, lowering the sap, and transporting it under cold conditions to the storage point. Implementing AI-based automation and leveraging its potential to perform the tasks of skilled tappers could ensure the timely availability of high-quality sap at a reasonable cost. In this context, automating one or more operations, such as massaging, slicing, or sap collection, would reduce the need for climbing, thereby ensuring a consistent supply of sap to meet the growing demand for health drinks and value-added products (Alander et al., 2013). Similarly, AI could be an effective technique for producing premium sugar from *neera*, where parameters like processing temperature, time, and endpoint can be controlled through automation and image processing. An automated chiller box could serve as an efficient alternative to the current coco sap chiller used during *neera* tapping, enhancing operational ease.

Furthermore, AI-based technologies for in-planta or in-palm sugar concentration methods could significantly boost the coconut sugar industry. Incorporating AI into the coconut sap concentration device can significantly enhance its efficiency and operability, particularly by enabling remote monitoring and control. AI algorithms can be used to predict sap flow rate variations based on historical data, environmental factors, and sap collection trends. This predictive capability allows the heating unit to adjust its operational parameters dynamically, ensuring optimal concentration levels are maintained without manual intervention. For remote operation, AI can integrate with IoT sensors placed in the heating chamber to continuously monitor flow rates, sap temperature, and concentration levels. The data can be relayed to an operator on the ground through a mobile or web interface, allowing real-time adjustments. For instance, if the flow rate drops from 125 ml h^{-1} to 84 ml h^{-1}, the system can automatically adapt the heating settings to maintain consistency, reducing energy waste and improving process efficiency. This AI-driven approach minimizes the need for manual oversight and ensures precise control over sap concentration from the ground level.

Machine learning algorithms can analyze big data on sap collection, environmental conditions, and palm health to identify patterns and make recommendations for improving yield and quality (Batz et al., 2023; Jung et al., 2020). AI models, such as Convolutional Neural Networks are trained on large datasets to recognize sap quality indicators such as colour and droplet size. Key training parameters - learning rate, batch size, and epochs - should be optimized, and real-time processing is supported by high-performance computing or edge solutions. Accuracy is assessed using high end metrics, with models regularly updated to ensure reliability and precision in sap quality monitoring.

AI can also predict the optimal time for tapping based on factors such as temperature, humidity, and palm maturity (Alander et al., 2013). It can also identify the most productive palms and suggest targeted plant growth interventions to maximize sap yield (Batz et al., 2023). Additionally, AI-driven systems can continuously learn and adapt to changing conditions, further refining the sap collection process over time (Batz et al., 2023).

The application of AI and ML in sap collection is still in its early stages, but the potential benefits are significant. These technologies can reduce the reliance on manual labour, improve the consistency and quality of the sap collected, and ultimately increase the profitability of coconut palm farming (Alander et al., 2013; Batz et al., 2023).

Implementing AI in high rainfall regions or tall palm trees like coconut plantations is a challenging task. Moisture can disrupt sensors, leading to inaccurate data and impacting AI performance. Devices such as cameras and drones are also prone to damage from rain, raising maintenance costs. Connectivity issues are common in remote areas, especially during rain, hindering real-time data transmission. Power supply can be unreliable due to cloud cover affecting solar-powered systems. Installing sensors on tall trees adds complexity, and environmental factors like wind and rain can distort readings. Additionally, the cost of durable, weather-proof equipment and reliable infrastructure and adoption of application of AI in such areas are challenging. Addressing these issues requires robust technology, strong data systems, and tailored models.

5. Decision Support Systems (DSS)

Decision Support Systems (DSS) are computer-based systems that help farmers and operators make informed decisions by analyzing data and providing recommendations (Bonfil et al., 2004; Gallardo et al., 2020; Kaklauskas, 2015). In coconut palm sap collection, DSS can be used to optimize the tapping process, monitor palm health, and predict sap yield (Bonfil et al., 2004). By integrating data from various sources, such as IoT devices, weather forecasts, and historical records, DSS can provide real-time insights that help improve the efficiency of sap collection (Bonfil et al., 2004; Lee and Jung 2020; Shaikh et al., 2022).

For example, a DSS could analyze data on weather conditions, palm health, and sap flow rates to recommend the optimal time for tapping (Bonfil et al., 2004; Shaikh et al., 2022). It could also suggest targeted interventions, such as adjusting the tapping technique or applying specific fertilizers, to improve yield and quality. By providing actionable insights, DSS can help farmers and operators make better decisions, reduce waste, and increase profitability (Bonfil et al., 2004).

6. Vision-Based AgroBots

Vision-based AgroBots represent an advanced approach to integrating machine vision with robotics for agricultural applications. These systems use cameras and image processing algorithms to gather and analyze visual data, enabling automated decision-making and task execution (Luo et al., 2023; Shashkova et al., 2022). In the context of coconut palm sap collection, Vision-Based AgroBots can offer significant improvements in efficiency and accuracy by automating tasks such as sap collection and quality assessment. It can be useful for larger area.

One key advantage of Vision-Based AgroBots is their ability to perform real-time monitoring of sap flow and quality (Siejak et al., 2024). By capturing images of the sap and analyzing its colour, clarity, and other visual attributes, these systems can ensure that only high-quality sap is collected. Additionally, Vision-Based AgroBots can be programmed to detect and address issues such as blockages or leaks in the collection system, reducing the risk of spoilage and waste (Shashkova et al., 2022). Moreover, Vision-Based AgroBots can facilitate more precise and targeted interventions (Starostin et al., 2023). For instance, these systems can identify individual palms that are underperforming and recommend specific actions to improve sap yield and quality (Shashkova et al., 2022). By automating these tasks, Vision-Based AgroBots can reduce the need for manual labour and improve overall productivity in sap collection.

7. Application of IoT and Big Data

The integration of IoT and Big Data into sap collection processes can provide valuable insights and enhance decision-making. IoT devices, such as sensors and smart meters, can monitor various parameters related to sap collection, including sap flow rate, temperature, and humidity (Ayaz et al., 2019). These devices transmit real-time data to a central system, where it can be analyzed to optimize sap collection and processing (Alahmad et al., 2023).

For example, IoT sensors can detect fluctuations in sap flow and trigger automated adjustments to the tapping process (Ayaz et al., 2019). Big Data analytics can be used to identify trends and patterns in sap collection data, such as optimal tapping times and conditions (Alahmad et al., 2023). This information can be used to make data-driven decisions, improve sap quality, and increase overall efficiency.

However, challenges such as the cost of IoT devices and the need for reliable internet connectivity in remote areas must be addressed to fully realize the potential of IoT and Big Data in sap collection (Alahmad et al., 2023). Despite these challenges, the integration of these technologies offers significant opportunities for improving sap collection processes and enhancing the quality of coconut palm sap (Ayaz et al., 2019).

Implementing IoT systems for monitoring coconut palm sap collection during the rainy season presents several challenges. High humidity and frequent precipitation can lead to issues with sensor performance and data accuracy. Moisture can interfere with sensor readings and cause malfunctions, leading to unreliable data. Additionally, water accumulation on sensors and other hardware can damage the equipment or degrade its functionality. Ensuring connectivity and stable data transmission becomes more difficult as rain can disrupt wireless signals and power supplies. Proper protective measures, such as weatherproof enclosures and robust calibration processes, are essential to mitigate these issues and maintain reliable IoT operations during adverse weather conditions. Also, operating IoT systems in high humidity areas, such as Kasaragod, Kerala, where the average humidity exceeds 80%, presents significant challenges. High humidity can lead to condensation on electronic components, causing short circuits, reduced lifespan, and operational failures. Sensors and communication devices may experience degraded performance due to moisture absorption, affecting data accuracy and reliability. Adapting IoT infrastructure to such conditions requires specialized materials and designs to mitigate these issues effectively.

8. Challenges and Future Directions

While advancements in technology offer promising solutions for improving coconut palm sap collection, several challenges remain to be addressed. The high cost of advanced technologies such as robotics, machine vision, and IoT devices, may limit their adoption, particularly among small-scale farmers (Batz et al., 2023). Additionally, the complexity of integrating these technologies into existing sap collection systems presents technical hurdles (Shashkova et al., 2022).

Future research and development efforts should focus on addressing these challenges by exploring cost-effective solutions and developing technologies that are compatible with traditional sap collection practices. Besides, efforts should be made to enhance the reliability and durability of these technologies, ensuring their effectiveness in diverse and challenging environments (Batz et al., 2023; Shashkova et al., 2022).

Conclusion

The collection of coconut palm sap has traditionally been a labour-intensive and challenging process, but advancements in technology are opening up new possibilities for improving efficiency, yield, and quality. Robotic systems, machine vision, AI, IoT, Big Data, and DSS are all playing a role in transforming the way sap is collected and processed. These technologies offer significant opportunities for reducing the reliance on manual labor, improving the consistency and quality of the sap collected, and ultimately increasing the profitability of coconut palm farming.

However, several challenges must be addressed to fully realize these benefits. The high cost of advanced technologies, the need for reliable power sources and internet connectivity in remote areas, and the complexity of data management are all potential barriers for widespread adoption. Nonetheless, the future of coconut palm sap collection is bright, with technology playing an increasingly important role in driving innovation and improving sustainability in the industry.

References

Alahmad, T., Neményi, M., Nyéki, A. (2023). Applying IoT sensors and big data to improve precision crop production: a review. *Agronomy* 13(10): 2603.

Alander, J.T., Bochko, V., Martinkauppi, B., Saranwong, S., Mantere, T. (2013). A review of optical non-destructive visual and near-infrared methods for food quality and safety. *Int J Spectroscopy* 2013(1): 341402.

Ayaz, M., Ammad-Uddin, M., Sharif, Z., Mansour, A., Aggoune, E.H.M. (2019). Internet-of-Things (IoT)-based smart agriculture: Toward making the fields talk. *IEEE Access,* 7: 129551-129583.

Batz, P., Will, T., Thiel, S., Ziesche, T.M., Joachim, C. (2023). From identification to forecasting: the potential of image recognition and artificial intelligence for aphid pest monitoring. *Frontiers in Plant Sci*14, 1150748. https://doi.org/10.3389/fpls.2023.1150748.

Beltran Ortega, J., Martinez Gila, D.M., Aguilera Puerto, D., Gamez Garcia, J., Gomez Ortega, J. (2016). Novel technologies for monitoring the in-line quality of virgin olive oil during manufacturing and storage. *J Sci Food Agric* 96(14): 4644-4662.

Bonfil, D.J., Karnieli, A., Raz, M., MufradiI, Asido S., Egozi H., Schmilovitch, Z. (2004). Decision support system for improving wheat grain quality in the Mediterranean area of Israel. *Field Crops Res* 89(1):153-163.

Dalibard, C. (1999). Overall view on the tradition of tapping palm trees and prospects for animal production. *Livestock Res Rural Dev11*(1): 1-37.

Ellsäßer, F., Röll, A., Ahongshangbam, J., Waite, P.A., Schuldt, B., Hölscher, D. (2020). Predicting tree sap flux and stomatal conductance from drone-recorded surface temperatures in a mixed agroforestry system - a machine learning approach. *Remote Sensing* 12(24):4070.

Ezzatzadegan, L., Yusof, R., Morad, N.A., Shabanzadeh, P., Muda, N.S., Borhani, T.N. (2021). Experimental and artificial intelligence modelling study of oil palm trunk sap fermentation. *Energies* 14(8):2137.

Fahey, T., Pham, H., Gardi, A., Sabatini, R., Stefanelli, D., Goodwin, I., Lamb, D.W. (2020). Active and passive electro-optical sensors for health assessment in food crops. Sensors, 21(1):171.

Fountas, S., Mylonas, N., Malounas, I., Rodias, E., Hellmann Santos, C., Pekkeriet, E. (2020). Agricultural robotics for field operations. *Sensors* 20(9):2672.

Gallardo, M., Elia, A., Thompson, R.B. (2020). Decision support systems and models for aiding irrigation and nutrient management of vegetable crops. *Agricultural Water Manage* 240, 106209.

Hai, A., Rambabu, K., Dhaheri, A.S., Kurup, S.S., Banat, F. (2024). Tapping into palm sap: Insights into extraction practices, quality profiles, fermentation chemistry, and preservation techniques. *Heliyon* 10(15), e35611, https://doi.org/10.1016/j.heliyon.2024.e35611.

Hebbar, K.B., Arivalagan, M., Manikantan, M.R., Mathew, A.C., Chowdappa, P. (2015). Coconut inflorescence sap and its value addition as sugar–collection techniques, yield, properties and market perspective. *Current Sci* 109(8): 1411-1417, doi:.10.18520/v109/i8/1411-1417.

Hebbar, K.B., Pandiselvam, R., Manikantan, M.R., Arivalagan, M., Beegum, P.P.S., Chowdappa, P. (2018). Palm sap—quality profiles, fermentation chemistry, and preservation methods. *Sugar Tech* 20:621–634 https://doi.org/10.1007/s12355-018-0597-z.

Jung, S.H., Kim, J.Y., Park, J., Huh, J.H., Sim, C.B. (2020). A study on acer mono sap integration management system based on energy harvesting electric device and sap big data analysis model. *Electronics* 9(11), 1979.

Kaklauskas, A. (2015). Intelligent decision support systems. In: *biometric and intelligent decision making support.* intelligent systems reference library, vol 81. Springer, Cham. https://doi.org/10.1007/978-3-319-13659-2_2 pp 31-85.

Lee, H.S., Jung, S.H. (2020). Analysis model evaluation based on IoT data and machine learning algorithm for prediction of acer mono sap liquid water. *J Korea Multimedia Soc* 23(10):1286-1295.

Luo, X., Sun, Q., Yang,T., He, K., Tang, X. (2023). Non-destructive determination of common indicators of beef for freshness assessment using airflow-three dimensional (3D) machine vision technique and machine learning. *J Food Eng* 340, 111305.

Natarajan, S., Ponnusamy, V. (2022). Agri-food products quality assessment methods. *Computer Vision and Machine Learning in Agriculture* Volume 2, pp121-136.

Okere EE, Arendse E, Nieuwoudt H, Fawole OA, Perold WJ, Opara U L (2021). Non-invasive methods for predicting the quality of processed horticultural food products, with emphasis on dried powders, juices and oils: A review. *Foods* 10(12), 3061.

Pandiselvam, R., Manikantan, M.R., Binu, S.M., Ramesh, S.V., Beegum, P.P.S., Gopal, M., Shil, S. (2021). Reaction kinetics of physico-chemical attributes in coconut inflorescence sap during fermentation. J Food SciTechnol58: 3589-3597.

Ranasinghe, C.S., Waidyanatha, U.D.S. (2003). Ethrel stimulation of inflorescence sap flow in tapped coconut (*Cocos nucifera*) palms. *Experimental Agric* 39(2): 161-166.

Rashid, M., Bari, B.S., Yusup, Y., Kamaruddin, M.A., Khan, N. (2021). A comprehensive review of crop yield prediction using machine learning approaches with special emphasis on palm oil yield prediction. *IEEE Access*, 9, 63406-63439.

Shaikh, F.K., Karim, S., Zeadally, S., Nebhen, J. (2022). Recent trends in internet-of-things-enabled sensor technologies for smart agriculture. *IEEE Internet of Things J* 9(23): 23583-23598.

Shashkova, I.G., Romanova, L.V., Kupriyanova, M.V., Cherkashina, L.V. (2022). The use of modern robotic systems in the agro-industrial complex. *IOP Conference Series: Earth and Environmental Science* 949, 012024, doi:10.1088/1755-1315/949/1/012024, 8 pages.

Siejak, P., Przybył, K., Masewicz, Ł., Walkowiak, K., Rezler, R., Baranowska, H.M. (2024). The prediction of pectin viscosity using machine learning based on physical characteristics—Case Study: Aglupectin HS-MR. *Sustainability* 16(14), 5877.

Starostin, I.A., Eshchin, A.V., Davydova, S.A. (2023) Global trends in the development of agricultural robotics. *IOP Conf. Series: Earth and Environmental Science* 1138, 012042, doi:10.1088/1755-1315/1138/1/012042, 8 pages.

Suganob, N.J., Bautista, M.G.A., Concepcion, R., Dadios, E. (2022) Coconut inflorescence sap collection for future food and bioresource technological advancements. In *2022 IEEE 14th International Conference on Humanoid, Nanotechnology, Information Technology, Communication and Control, Environment, and Management* (HNICEM) (pp. 1-6). IEEE.

Suganob, N.J.O., Concepcion, R., Vicerra, R.R.P. (2023). Adaptive neuro-fuzzy regression of coconut inflorescence texture phenotypes through spectral analysis using computer vision. In *2023 8th International Conference on Business and Industrial Research* (ICBIR) (pp. 561-567). IEEE.

Chapter 6

Fermentation Chemistry, Preservation Methods, Storage and Shelf Life of Coconut Palm Sap

G. Murali*
S. V. Ramesh†
V. Krishnakumar
K. Sharanya
M. Asha
M. R. Manikantan
and K. B. Hebbar
ICAR-Central Plantation Crops Research Institute, Kasaragod, Kerala, India

Abstract

Coconut palm sap, extracted from the coconut tree, is a sweet, nutrient-rich liquid that undergoes rapid fermentation due to its high sugar content and natural microbial presence. The sap contains primarily sucrose, along with glucose and fructose. Immediately after harvest, microorganisms such as yeast, lactic acid bacteria, and acetic acid bacteria begin to breakdown the sugars. The fermentation process occurs in two main stages: alcoholic and acidic fermentation. This fermentation process is swift, typically starting within a few hours, and can rapidly turn the sap into a mildly alcoholic beverage (toddy) or vinegar if left unrestricted.

* Corresponding Author's Email: mgcpcri@yahoo.co.in
† Corresponding Author's Email: ramesh.sv@icar.gov.in

In: Coconut Palm Sap
Editors: K. B. Hebbar, V. Krishnakumar and S. V. Ramesh
ISBN: 979-8-89530-640-6

Due to the rapid fermentation of coconut palm sap, various preservation techniques such as refrigeration, pasteurization, chemical preservatives, drying or concentrating and fermentation control techniques are used to prevent spoilage and prolong its shelf life. While methods and the use of preservatives can extend its shelf life, fermentation can be controlled to produce stable byproducts like coconut syrup, vinegar or alcohol. This chapter provides an over view of the chemical and microbial process of fermentation, as well as various preservation methods employed to enhance the shelf life of coconut palm sap.

Keywords: fermentation dynamics, coconut sap, microbiome, probiotics, palm sap shelf life

1. Introduction

Coconut inflorescence sap or palm sap is relished in many regions around the world and is made by collecting sap from palm trees. Palm sap is rich in essential minerals such as sodium (Na), potassium (K), calcium (Ca), and iron (Fe), and also contains ascorbic acid and polyphenols. Due to its antioxidant properties, palm sap can be considered both nutritious and healthy. Traditionally, palm sap has been used as folk medicine, offering benefits like improved vision and gastrointestinal health. The sap is high in sugar and minerals, which creates an ideal environment for the growth of yeast and lactic acid bacteria (LAB). Fermentation begins soon after collection, and within a few hours, the alcohol content of the sap can reach up to 4-5% (Figure 1).

(a) (b)

Figure 1. Fresh (a) and fermented coconut palm sap (b).

2. Tapping for Sap Production

Coconut palm trees can be tapped for sap from a young age. During the harvesting process, one to two millimeters of the spadix is sliced off each time for sap collection. This method can be continued for 40–45 days until the spadix is reduced to a stump. A coconut palm can be tapped for approximately 20 years (Hebbar et al., 2015; Levang, 1988). Tapping inflorescence sap is a critical step, as the sap is prone to immediate degradation, which can adversely affect the quality of the palm sugar produced from it (Kurniawan et al., 2018). The tapping process for coconut palms, as practiced in coastal Kenya was described by Kadere et al., (2004). The procedure begins with tappers forcing the emerging inflorescence into a pendant position by notching the woody bract surrounding it and tying the inflorescence in place with rope. Once the inflorescence reaches its full size, the tip is cut off, and the branches are bound together in a bundle to allow sap to exude into a collection vessel. This vessel is emptied two to three times a day, and the cut end of the inflorescence is shaved daily to ensure a continuous flow of sap. Importantly, there is no beating or bruising of the inflorescence before tapping. Among the fourteen different tapping methods evaluated in Central Java, Indonesia, five methods were found to be the best for tapping and preserving coconut sap, which can produce crystalline coconut sugar with sucrose, reducing sugar, moisture, and ash content that meet the SNI 3743-2021 standards (Mustaufik et al., 2022). The details are given in Table 1.

Chinnamma et al., (2019) implemented a scientific approach for harvesting sap by modifying the traditional tapping methods used by toddy tappers (Figure 2). This method aims to enhance the quality and yield of the harvested sap. The main steps are:

1. Selection of Healthy Palms: Choosing robust and disease-free coconut palms for tapping.
2. Surface Sterilization: Sterilizing the crown of the palm to prevent contamination.
3. Cleaning of the Spadix: Ensuring the spadix is clean before tapping begins.
4. Beating of the Spadix:
    - Initial beating from the base to the top for 3–4 days to stimulate sap flow.
    - Chopping the tip of the spadix with a sharp, sterile knife.

- Continuing the beating twice daily (morning and evening) for an additional 10–15 days.

5. Covering the Chopped End: Daily covering of the chopped spadix end with a sterilized plastic mesh to maintain hygiene.
6. Application of Sterilized Clay: Applying sterilized clay to the cut end of the spadix to seal and protect it.
7. Insertion of Sterilized Plastic Vessel: After 15-20 days of beating, inserting a sterilized plastic vessel with an anti-fermentation solution into the spadix, depending on the sap flow observed.
8. Collection of Sap: Collecting the sap twice daily, in the morning and evening.

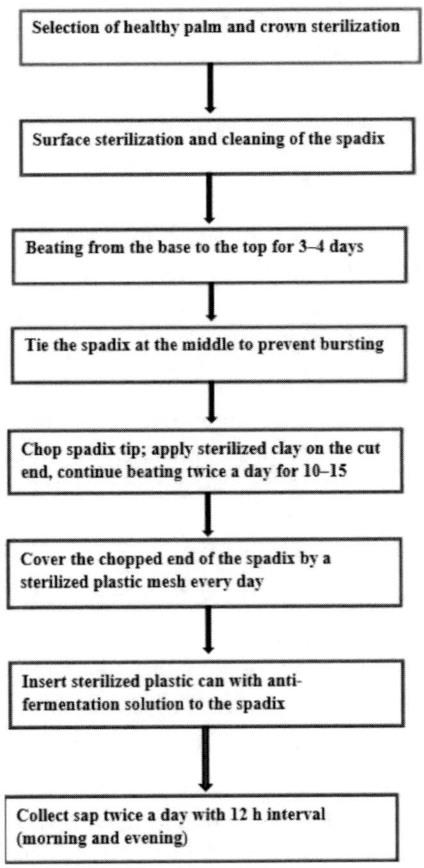

Figure 2. The modified harvesting steps of coconut sap (Chinnamma et al., 2019).

Table 1. Chemical characteristics of coconut sap due to the best tapping method and storing coconut sap (Mustaufik et al., 2022)

Tapping and storing of coconut sap	Chemical characteristics of coconut sap		
	pH	Total dissolved solids (°Bx)	Sucrose content (%)
Method I: tapping time from morning to noon for 6 hrs (06.00 AM-12.00 PM), without laru, heating, sheltered duration or directly produced into crystal coconut sugar.	5.56	16.23	15.93
Method V: tapping time from morning to noon for 8 hrs (06.00 AM-02.00 PM), + 2% laru, without heating, sheltered duration or directly produced into crystal coconut sugar.	7.14	18.49	17.29
Method VI: tapping time from morning to noon for 8 hrs (06.00 AM-02.00 PM), + 2% laru, without heating, sheltered duration for 3 hrs and then produced into crystal coconut sugar	6.39	17.82	16.60
Method VII: tapping time from afternoon to night for 8 hrs (02.00 PM-10.00 PM), + 2% fresh, without heating, without shelter duration or directly produced into crystal coconut sugar.	8.13	18.72	17.52
Method VIII: tapping time from afternoon to night for 8 hrs (02.00 PM-10.00 PM), + 2% laru, without heating, shelter duration for 3 hrs, then produced into crystal coconut sugar.	7.33	17.43	16.93

The freshly tapped sap must be processed immediately to prevent fermentation, which can significantly alter its taste and quality. Fermented sap develops a pungent odor and acidic taste due to increased acid content and the formation of ethyl esters and ethyl alcohol (Borse et al., 2007). Traditional tappers often use the lime coating on the inner lining of the collection pot to prevent fermentation. Although several other methods have been attempted to inhibit fermentation, they are less effective (Hebbar et al., 2018). Preventive measures during harvesting, processing, and storage are crucial, as the sap is highly susceptible to rapid fermentation. Due to the challenges of preserving the sap, it is often converted into value-added products and marketed globally on a commercial scale.

Saccharomyces cerevisiae and *Zymomonas mobilis* are the primary microorganisms involved in the fermentation process of palm sap to produce alcohol. *S.cerevisiae* is naturally present in palm sap and is crucial for spontaneous fermentation. This yeast converts the sugary sap into ethanol, lactic acid, and other volatile compounds. In addition to *S.cerevisiae*, palm wine also contains various probiotic organisms, including *Bacillus, Saccharomyces, Lactobacillus, Streptococcus, Leuconostoc.* These probiotics contribute to the production of vitamins and digestive enzymes, and they help enhance the immune system. In rural regions of India and the Terai area of Nepal, palm sap and palm wine are consumed for their health benefits, such as improving eyesight and gastrointestinal health.

3. Sap Quality

Ensuring the safety and quality of coconut sap requires meticulous pretreatment of collecting containers (Borse et al.2007). Microbial growth, encompassing bacteria, yeast and mold poses significant threat to coconut sap quality during harvesting and transportation, emphasizing the necessity for temperature reduction and adherence to hygienic practices (Ghosh et al.2018; Ysidor et al., 2015). Okafor (1975) and Sanni (1993) reported that yeasts particularly *S. cerevisiae* are dominating among fermentation organisms. However, the shelf life of coconut sap is always a limiting factor for its valorization.

The challenge in maintaining the quality of the coconut sap beverage lies in the quality of incoming sap from the farms. Identifying factors influencing post-harvest sap quality is crucial for fostering mutual benefits, with parameters such as pH, total soluble solids (TSS), and bubble formation serving as key indicators for farmers in the coconut fields (Hebbar et al., 2015; Kapilan et al., 2015; Saidan et al., 2020). The pH of coconut sap continues to decrease during collection due to increasing microbial loads that leads to increasing total acidity of the sap (Iwuoha and Eke 1996). The decrease in pH causes sucrose inversion and lowers the quality of sap. Thus, the long collection time causes lowering of pH, high amounts of total acidity and reducing sugar resulting in the low recovery of sugar from sap (Naknean et al.2010). Hebbar et al., (2015) indicated that fresh sap is slightly alkaline in nature with pH of 7.5-8.0 and varies from palm to palm. The fermentation commences within two to three hours of collection under ambient conditions showing a decreasing pH trend, consequently turning it into completely a fermented sap with pH of 3.5. Fresh sap with pH around 7.5 contains about 15% sugar that reduces to 6% at pH level of 4.0. Storing the sap in a deep freezer (-2°C to -1°C) maintains the initial pH.

The fast spoilage of coconut palm sap is because of its rich nutritional composition and various microorganisms associated with unhygienic production conditions and environment. The glaucids of the sap undergo considerable deterioration after harvesting producing many volatile compounds like acids, alcohols, ketones, aldehydes and esters which cause instability of the product (Borse et al., 2007). Compounds like isoamyl alcohol along with other compounds such as ethyl caprate, ethyl lactate, ethyl caprylate, phenyl ethyl alcohol, palmitoleic acid and dodecanoic acid are responsible for odor of fermented coconut sap.

Spontaneous deterioration of sap due to its highly fermentable nature has prompted development of newer sap processing technologies to improve its shelf life as clarified and processed sap is suitable for longer storage (Ramalakshmi et al., 2004). When collected in controlled containers for 10–12 h, unfermented coconut sap exhibits a slightly alkaline, golden-brown color with a mild sweetness. Conversely, traditional sap collection without temperature control leads to microbial fermentation, resulting in an acidic, off-white sap with a distinct fermented odor (Hebbar et al., 2015; Purnomo 2007).

The sap that has undergone fermentation cannot be processed and moulded into the good palm sugar. Poor quality of sap produces sour taste, and hence, the sap remains unpalatable. Adding ginger can improve flavour quality, and based on physico-chemical parameters, addition of red ginger at a concentration 10% is considered a best option (Febriyanti et al., 2015). Ysidor et al., (2015) reported that the quality of coconut sap deteriorates during storage, especially the pH and sugar contents drop significantly, while the phenol content increases to more than double.

The inherent presence of yeast in the sap enhances natural fermentation. To prevent the fermentation process, the yeast load in fresh sap should be reduced at the earliest possible. Electrospun polycaprolactone nanofibrous membrane can be used for the removal of yeast from coconut sap (Leena et al., 2021). The filtered sap had significant changes in titratable acidity, pH, and colour. The content of TSS did not change; however, total polyphenolic content and minerals showed slight reductions. The sap filtrate obtained through the optimized nanofibrous membrane showed a 2 log-reduction in yeast load, thereby suggesting potential of the electrospun nanofibrous membrane for *neera* filtration, quality retention, and shelf-life extension applications.

Analysis of key factors affecting coconut sap quality after harvesting in Thailand (Wiboonsirikul et al., 2024) indicated that storage and transportation temperatures, and the type of preservative used impacted the pH of the sap. However, the washing method and type of preservative used did not affect the TSS. The researchers recommended use of payom wood (*Shorea roxburghii* G. Don) as a preservative instead of chemicals, faster collection of sap within four hours, coupled with stringent temperature control during transportation

(≤10°C) to ensure optimal quality. Besides, use of clean water, thorough rinsing, and drying of containers are crucial practices to ensure the quality of coconut sap.

4. Storage and Sap Quality

The Defence Food Research Laboratory (DFRL), India in collaboration with the Central Food Technological Research Institute (CFTRI), Mysore, India has developed a process for the preservation of coconut sap. This consists of collecting the coconut sap early in the morning in a hygienic condition, transporting immediately in a chilled condition (4°C), clarifying the sap, adjusting the pH of the sap by adding an acidulant citric acid in the range of 0.04 to 0.2%, blending with certain additives (nisin @ 5 to 15 ppm) followed by filling in glass bottles, sealing with crowns and pasteurization at 90-95°C for 3 to 7 min to enable the sap to retain its natural constituents. Heat preservation methods such as pasteurization and sterilization are necessary to preserve and extend the shelf life of the product. Thermal stress could be reduced by adding bio-preservative nisin at a minimum concentration of 10 ppm both for in-pack pasteurization and retort processing. Furthermore, as the sap is highly photosensitive due to the presence of ascorbic acid, foil-based multi-layer packaging materials such as polyethylene terephthalate (PET) or aluminium foil or cast polypropylene film were found suitable to prevent the adverse effect of light. In order to increase the product appeal, the suspended particles in the sap are to be removed by centrifugation at 4000 rpm for 10 min. The product as such without any additive treatment as well as heat processing was stable up to 3 days. When the product was processed either by in-pack pasteurization or through retort pouch processing, the shelf life could be about one year under refrigerated condition and 30 days under ambient conditions. The quality of sap was similar under both the cases of retort processing or in-pack pasteurized samples; however, latter will be more cost-effective (Jayashree 2013).

There will be reduction in pH if the sap is stored for longer durations. Ramalakshmi et al., (2018) noticed reduction of pH from 7.3 to 5.9 during the incubation period of 4 h followed by the drastic reduction to pH 3.3. The fresh coconut sap contains 14.35, 0.32 and 0.28% of sucrose, fructose, and glucose, respectively. However, within the first 12 h of storage, sucrose content decreases up to 1.04% and fructose and glucose content increases to 5.75 and 8.72%, respectively. Considerable increase in the alcohol content, from

0.013% at 0th hour to 5.17% in the 40th hour of storage was documented. High amount of microbial load (total plate count and yeast and mold count) was also seen in the sap samples.

Generally, *neera* is preserved using a combination of processing methods such as pasteurization, filtration, use of chemical or bio-preservatives, clarifying agents, centrifugation and carbonation. However, various technologies are not effective in arresting the growth of micro-organisms in *neera* and the treatments change its organoleptic qualities (Hebbar et al., 2018). Even at refrigerated temperatures, the microbes maintain their viability and spoil the quality of *neera* within 72 h of storage and the time to spoilage may vary according to the initial microbial population. The survival of microbes at these low temperatures is due to the expression of various proteins, such as cold-inducible protein, cold shock protein and aquaporin, which extensively regulates the metabolic function of micro-organisms (Aguilera et al., 2007). The non-aqueous compounds of food, such as carbohydrates, lipids, protein and organic acids, change during storage and affect the pH, titratable acidity and ionic strength of the food (Damiani et al., 2013). Moreover, the stability depends on the type, chemical composition and microbial population of food.

Technologies have been standardized for arresting fermentation of *neera* and its preservation as alcohol free product (*Keraamritham*) under refrigerated condition with a keeping quality of more than three months (Naik et al., 2013). The nutritional status of *Keraamritham* is presented in Table 2.

Thermal pasteurization for 5 min at 90°C caused degradation of its natural colour and imparted a cooking flavour. The percent titratable acidity and reducing sugar were also low under this treatment. Pasteurization at 80°C for 5 min maintained the pH throughout the storage period of 16 days, whereas there was drastic decrease of pH under control treatment due to microbial spoilage. The treatment at 60°C was not effective in inhibiting the fermenting microorganisms during the entire storage period. The intermediate thermal treatments of 70°C and 80°C minimized the degradation of the sap and maintained optimal biochemical constituents of the sap (Prashanth and Patil 2020).

Frozen storage of coconut sap at -20°C caused better inhibition of lactic acid bacteria, which enabled preservation of the sap from the metabolic conversion (Sukumaran and Radhakrishnan 2021). The physicochemical qualities were also retained for a longer period when stored at -20°C than at atmospheric storage. Colour, viscosity, total protein and ethanol showed minimal alterations from the control and better sensory attribute up to 21 days at -20°C. At the end of 28 days, the sap exhibited a pH of 5.37, TSS of 14.0 °Brix and total acidity of 0.82 mg l^{-1} at -20°C. The Indian patent received by Bashyam et al., (2004) for an invention relates to a process for the preservation of de-odourised coconut sap. The process, in particular, describes the method of preservation of coconut sap in glass bottles.

5. Fermentation of Palm Sap

The sweetness of coconut sap is composed mostly of carbohydrates, such as sucrose, glucose, and fructose, among other nutrients. Bacteria, along with the high concentration of these simple sugars, as a bioavailable carbon source, and the non-hermetic state of the coconut sap tapping, create ideal conditions for quick fermentation. Coconut sap is highly prone to fermentation due to its diverse microflora, which includes yeasts, bacteria, fungi, and degrading enzymes. Its high sugar content attracts various microorganisms, particularly aerobic mesophiles (Nwachukwu et al., 2006; Ogbulie et al., 2007). Monitoring parameters like pH, titrable acidity, volatile acidity, reducing sugar, TSS, alcohol content, ester content, and aldehyde content during the fermentation phase and storage have commercial and economic consequences for the future exploitation of the palm. Other than the parameters suggested above, analysis of amino acids, total nitrogen, volatiles, vitamin C, and phenolic acids provide additional information on the course of fermentation.

The natural fermentation of coconut sap involves initial lactic acid fermentation, followed by alcoholic fermentation and, finally, acetic acid fermentation. The entire process takes place due to several isolates of yeasts and bacteria (Atputharajah et al., 1986; Tomomatsu et al., 1996; Xia et al., 2011; Ysidor et al., 2015). Vidanapathirana et al., (1983) reported that coconut sap fermentation occursin three phases, the first of which is predominated by fermentation caused by *Leuconostoc, Lactobacilli, Streptococci, Bacilli,* and *Enterobacter* results in a pH reduction from 7.0 to 4.0, without any alcohol production. Yeasts (*Saccharomyces chevalieri*), Pichia (two strains), and Candida (two strains) dominated the second phase (30 h–5 days), which resulted in an increase in alcohol, without a change in pH being maintained at 4.0. *S. chevalieri* was the main alcohol producing organism in the sap during this phase. The third phase, which is a slow one, commences after two to three weeks later, dominated by acetic acid bacteria.

Table 2. Nutritional qualities of Keraamritham (Naik et al., 2013)

Parameters		Minerals (ppm)	
Total sugar	15-18%	P	90.5
Vitamin C	1.3 mg 100 ml^{-1}	K	60.0
pH	6.8	Ca	15.0
Acidity	10.0 m eql^{-1}	Fe	45.0
Total poly phenols	8.0 mg 100 ml^{-1}	Na	9.5

Coconut inflorescence sap is rapidly fermented by a heterogeneous population of microorganisms that originate from the environment of the coconut inflorescence. Wijesinghe and Samarajeewa (1988) found that more than 50% of the microorganisms isolated from fermenting coconut inflorescence sap could produce ≥ 9% (v/v) ethanol in batch cultivation using sap containing 15-18% (w/v) sucrose. Wellala et al., (2004) isolated 12 yeasts from naturally fermenting sap in Sri Lanka and reported that all isolates were identified as members of the *Saccharomyces* genus.

Borse et al., (2007) studied the chemical composition of volatiles from fresh and fermented coconut sap and the effects of processing on these compositions. They have reported 21 compounds, constituting more than 98% of the volatiles from fresh sap, and 12 compounds, representing more than 95% of the volatiles in fermented sap. Bacteria induce lactic acid fermentation (first stage of fermentation) of the sap, which brings down the neutral pH to acidic (from 7 to 5) in 8-20 h and promotes the activity of yeasts, mainly *Saccharomyces.* The yeasts convert sucrose into invert sugar, and the ensuing alcoholic fermentation produces ethanol. The generation of ethanol provides a favourable environment for the acetic acid bacteria, and acetic acid fermentation causes a further decrease in the pH (from 4 to 3).

By the end of fermentation, acetic acid bacteria develop in two to three days, and when toddy is allowed to distil, it yields a spirit known as arrack (Atputharajah et al., 1986). Toddy and arrack are mostly made from coconut sap, which suppresses non-ethanol-generating bacteria and permits pure yeast cultures to synthesize ethanol to variable degrees (Sánchez and Ríos 2002). The protein content in the sap also changes due to fermentation. Coconut sap is fermented primarily by yeast of the genus *Saccharomyces cereviceae* that convert glucose to ethanol and *Acetobacter* sp. bacteria that oxidize ethanol to acetic acid during the tapping and storage processes (Xia et al., 2011; Setyawan and Ninsix 2016).

The by-products of fermentative microorganisms or CO_2 turn the sap more turbid. Due to natural fermentation, the sugars in the sap (12-18%) are converted to alcohol and, further, to acids and other volatile compounds. Ethanol is the main alcohol present in toddy, and in two days, its content reaches its maximum. In the early stages of coconut sap fermentation, *Schizo saccharomyces pombe* ferments sucrose and glucose, but in the later stages *Pichia angophorae* ferments sucrose, and *Kloeckera apiculata,* and *Candida glabrata* ferment fructose and galactose (Kalaiyarasi et al., 2013). Saini et al., (2021) obtained nine bacterial isolates from coconut sap, which were identified as *Serratia* sp. *Staphylococcus* sp, *Streptococcus* sp. *Bacillus* sp., *Microbacterium* sp., *Enterobacter cloacae, Klebsiella aerogenes*, *Bacillus cereus* and *Pantoea anthophila.*

While studying the fermentation kinetics of palm sap, Shetty et al., (2017) found that there was predominance of LAB in the sap until 7 h of incubation, whose count decreased thereafter with simultaneous increase in yeast population. Increase in yeast population was accompanied by increasing concentration of invertase and concomitant increase in the reducing sugars. Alcohol content at 13 h of fermentation was found to be around 4.1%. Pandiselvam et al., (2021) investigated the fermentation characteristics of sap (Kalparasa) at every 60 min. interval under ambient (31 ± 2°C) and refrigerated (5 ± 1°C) storage conditions. The pH and total sugar content of the sap declined rapidly under the former condition than the latter. The acidity, turbidity, and reducing sugar content increased in the sap stored under ambient condition. The reaction rate constant (k) of the vitamin C and total sugar degradation increased with the atmospheric fermentation. Despite the extensive usage of *Saccharomyces cerevisiae* in the alcoholic fermentation business, the major alcohol producing microbe in coconut sap is *Saccharomyces chevalieri* (Mat et al., 2022).

6. Total Soluble Solids

Total soluble solids (TSS) are the concentration of dissolved sugar, organic acids, soluble proteins, and amino acids in a fruit or fruit juice. A change in this parameter is a clear indication of a change in the sugar content of the product. Hebbar et al., (2018) reported that the TSS of coconut palm sap decreased to 6% at pH 4 from 15% at pH 7.5(freshly tapped palm sap). In the coconut palm wine, a significant drop in the TSS was observed from 21.5°Brix to 11.13°Brix in 6 days of fermentation. There was only a gradual decrease afterward and no further reduction in the sugar content after a Brix value of 10 (Khadka et al., 2024).

Ysidor et al., (2014) reported the changes in the TSS contents of four different coconut varieties during 120 h of storage at ambient temperature. TSS was found reduced for all the varieties with means varying from 16.31 ± 0.55-14.65 ± 0.41% to 7.8 ± 0.61-3.82 ± 0.89%. The change was slow in the first 12 h of storage for all the varieties and there was a rapid change from 24th hr to 96th hr and TSS remained almost constant until 120 h. During the earlier storage (0 h to 12 h), total carbohydrate contents (TCC) ranged between 16.08 ± 0.75 to 12.85 ± 0.62 g 100 $ml^{-1}$for all the coconut varieties. From 24th h (12.11 ± 0.79 to 12.58 ± 0.36 g 100 ml^{-1}), the TCC declined significantly until the 96th h of storage with means ranging between 1.26 ± 0.10 and 1.82 ± 0.02 g 100 ml^{-1}.

7. Reducing Sugar

Xia et al., (2011) reported a drop in total sugar content during the fermentation of fresh coconut inflorescence sap and an increase in reducing sugar till 3rdday of fermentation and a quick drop afterward. This increase in reducing sugar content could be due conversion of sucrose to glucose and fructose at the initial stage of fermentation because of the activity of microorganisms as the fermentation progresses. Hebbar et al., (2018) also observed a similar trend in coconut palm sap, there was an increase in the total reducing sugar and a reduction in total sugar during the early stages of the fermentation process. This is in contrast with the findings of Khadka et al., (2024) that the reducing sugar of the naturally fermented sap reduces over time at ambient temperature. A possible explanation for this could be that the microbial consumption rate of reducing sugar is constantly higher than the inversion rate of sucrose resulting in a net reduction in the quantity of sucrose.

Ysidor et al., (2014) reported reducing sugar content between 2.41 ± 1.30 and 3.73 ± 2.87 g 100 $ml^{-1}$for palm sap from four varieties of coconut palm at the beginning of storage. After 4 h these values decreased and ranged from 0.56 ± 0.37 to 2.45 ± 0.30 g 100 ml^{-1}. The values increased until 24 h of storage and reached maximum values between 6.6 ± 2.49 and 8.29±2.08 g 100 ml^{-1}. After this period, the reducing sugar dropped significantly until it reached values ranging from 0.07 ± 0.06 to 0.15 ± 0.12 g 100 ml^{-1}.

8. Total Acidity

The freshly collected coconut palm sap starts fermenting within 2-3 h at ambient temperature, and the pH can drop to 3.5 from its initial neutral pH (Hebbar et al., 2018). This decline in pH is associated with an increase in the total acidity. Xia et al., (2011) also reported an increase in total acidity from 1st day to 3rd day of the fermentation of coconut palm sap. This increase could be due to lactic acid fermentation. The rate of increase slowed on 4th and 5th day as alcohol fermentation began. There was a sharp rise in pH after the 5th day, which could be attributed to acetic acid fermentation. Ysidor et al., (2014) investigated the changes in the pH and total titratable acidities of the sap from four coconut varieties during storage at 35+/-2°C. The total titratable acidity increased from 0.26 ± 0.04 to 1.85 ± 0.09 g 100 g^{-1} during the 120 h of storage, with relative stability stages. Statistically significant differences were observed between the varieties. The pH of each coconut sap sample decreased from 7.32 ± 0.17 to 4.17 ± 0.15 during the 120 h of storage. The pH remained constant for the first 8 h of storage, and started declining until 96 h, to a value of 4.37 ± 0.06 to 4.47 ± 0.21. After 96 h, the pH remained nearly constant for all the varieties, for the rest of the storage period.

Eze and Ogan (1988) reported a steady decrease in pH from the near-neutral value of 6.6 in pure sap at the beginning of the fermentation to a stable value of 3.40-3.30 at 48 h. Titratable acidity, representing total dissociable protons in the sap, increased steadily from 2.1 mM in pure palm sap to 246.7 mM at 96th h of fermentation under these conditions. The stabilization of pH at around pH 3.30 after 48 h could be due to the efficient buffering of protons by the weak organic acids produced.

9. Volatile Acidity

Another measurable parameter associated with the fermentation of palm sap is volatile acidity. In fermented palm sap, acetic acid is the major volatile acid observed (Atputharajah et al., 1986). Xia et al., (2011) reported the changes in volatile acid content in coconut palm sap during fermentation. Volatile acidity remained unchanged for the first 4 days of fermentation and thereafter increased to a maximum value on 11th day of fermentation. This could be due to the onset of the acetic acid fermentation phase, which begins on 5^{th}day of fermentation.

10. Total Alcohol Content

Another parameter that could be analyzed during the fermentation process is alcohol by volume (ABV). Xia et al., (2011) reported that in coconut palm wine, production of alcohol was higher on 7th day (i.e., 9%), and decreased to about 8% on the 11th day of fermentation. Borse et al., (2007) identified the presence of isoamyl alcohol in fermented coconut palm sap (fermented for 24 h at 30 ± 2°C), along with elevated amounts of phenyl ethyl alcohol. The ethanol content increased from 0.07% to 2.56% v/v after 24 h fermentation at 30 ± 2°C. The fermented odour observed in palm sap could be due to the presence of isoamyl alcohol (Jackson and Linskens 2002).

11. Carbohydrate

Generally, palm sap contains mostly sucrose, followed by glucose and fructose. The sugar profile of palm sap varies depending on factors such as its geographical origin, season, extraction methods, and sample age. The total sugar content of coconut sap is high (16.19 g 100 ml^{-1}), compared to its reducing sugar content (0.68). The extent of the sap's freshness can be easily measured by the composition of total and reducing sugars. If the sap is fresh and unfermented, the reducing sugar content should be less than 1.0 g 100 ml^{-1} (Hebbar et al., 2013).

12. Ester and Aldehyde Contents

Ester content is expressed in terms of ethyl acetate. Esters are produced through microbial action by combining acids and alcohol via the esterification reaction. Some esters may contribute to bitterness, serve as flavour precursors, or participate in reactions with phenolics. The amount of ester becomes significant 4 days after the initialization of fermentation. In the coconut palm wine with natural fermentation, the amount of ester was similar for days 4, 10, and 12 and the days 6 and 8 were significantly different from other days. In the palm wine with pure culture added, the total amount of ester decreased significantly after the fermentation was completed (Khadka et al., 2024).

Aldehyde is one of the biochemical products formed as a part of fermentation and the conversion of methanol and ethanol to their respective aldehydes is the activity of alcohol dehydrogenase enzymes.

13. Phenolic Compounds

Phenolic compounds are secondary metabolites produced during plant metabolism and are much valued in the food industry because of their nutritional, antioxidant, and sensory properties. Phenolic acids are found in both free and bound forms in plants. Phenolic acids are in bound form when they form glycosidic bonds with sugar, proteins, and starch present in the plant matrix.

The total phenolic content increased slowly from 1^{st}day to 2nd day of fermentation and increased rapidly to a peak value at 58 h. The increase in phenolic content could be due to the release of bound phenolic acids by the acids produced during natural fermentation. After recording a peak value in the content at 58th h, phenolic acids showed no obvious change in their concentration. Xia et al., (2011) also identified five phenolic acids namely caffeic acid, gallic acid, p-coumaric acid, protocatechuic acid, and galanginin both in fresh sap and fermented sap. All five phenolic acids were higher in fermented coconut palm sap than in fresh coconut palm sap. Gallic acid was the dominant phenolic compound in fresh sap, around 350 μg l^{-1}, whereas caffeic acid content (730 μg l^{-1}) was the most abundant in fermented sap.

Hebbar et al., (2020) reported a phenolic content of 21.47+/- 0.48 mg GAE100 ml^{-1}coconut sap collected using a coco-sap chiller. LC-MS-MS analysis of the sap confirmed the presence of phenolic acids such as vanillic acid, syringic acid, ferulic acid, caffeic acid, gallic acid, p-coumaric acid,o-coumaric acid, 2,4-dihydroxy benzoic acid, gentisic acid, protocatechuic acid, trans-cinnamic acid, p-hydroxy benzoic acid and salicylic acid.

Ysidor et al., (2014) studied the change in phenolic acid content in different coconut palm varieties, MYD, WCT, PB 121+, PB 113+, during storage. During 0 h to 24 h of storage, the total phenolic content (TPC) remained constant for all the varieties, MYD (254.20 ± 3.40 to 331.91 ± 25.76 mg l^{-1}), WAT (565.80±2.37 to 696.60 ± 32.26 mg l^{-1}), PB 121+ (464.40 ± 5.91 to 636.75±28.02 mg l^{-1}) and PB 113+ (402.8 ± 7.87 to 517.09 ± 137.55 mg l^{-1}). Between 24 and 48 h, the values increased significantly and reached a maximum of 652.42 ± 72.69 mg l^{-1}(MYD), 1056.98 ± 229.59 mg l^{-1}(WAT), 1045.58 ± 276.47 mg l^{-1}(PB 121+) and 1316.24 ± 51.81 mg l^{-1} (PB 113+). At

96 h of storage, values ranged from 292.02 ± 25.04 mg $l^{-1}$to 601.14 ± 27.81 mg l^{-1}among the varieties. Throughout storage, the variety WAT showed the highest amount of total polyphenols, while MYD recorded the lowest. The decline in phenolic acid content after 48 h could be the result of the accelerated activity of enzymes, polyphenol oxidase, and phenolase at 35+/-2°C. There was no obvious change in the phenolic acid content after 56 h in a similar experiment by Xia et al., (2011). This difference could be due to different temperature conditions (25+/-2°C) in this experiment (Ysidor et al., 2014).

14. Amino Acid/Protein Content

Fresh coconut palm sap contains 901 mg 100 g^{-1} free amino acids as reported by Hebbar et al., (2018). Hebbar et al., (2020) found that fresh coconut inflorescence sap contains amino acids like alanine, serine, proline, valine, threonine, leucine asparagine, asparticacid, lysine, methionine, histidine, phenylalanine, arginine, tyrosine, tryptophan glutamic acid and citrulline,3,4-Dihydroxy-phenylalanine. Xia et al., (2011) reported the amino acid content of fresh coconut palm sap as 2.6 g kg^{-1}, which decreased on the first day of fermentation and there was no change on 2ndday, this could be due to the degradation of protein in the palm sap. There was a decrease in the amino acid content on the 3rdday and again an insignificant decrease later. This could be due to the production of amino acids by death or yeast autolysis(Penn et al., 1983).

Ysidor et al., (2014) compared the change in protein content of coconut palm saps of four varieties of coconut, during 120 h storage period. The protein contents of coconut palm sap varied between 0.14 ± 0.05 and 2.84 ± 0.30 g 100 ml^{-1} without any significant difference between the varieties. The protein contents remained stable until the 8th h of storage, from 0.35 ± 0.05 to 0.83 ± 0.66 g 100 ml^{-1} (MYD), 0.42 ± 0.01 to 0.59 ± 0.25 g 100 ml^{-1} (WAT), 0.49 ± 0.04 to 0.66 ± 0.43 g 100 ml^{-1} (PB 121+) and 0.38 ± 0.06 to 0.55 ± 0.30 g 100 ml^{-1} (PB 113+). From 8 h they increased significantly until 24 h and reached peak values ranging from 2.34 ± 0.37 to 2.84 ± 0.30 g 100 ml^{-1} for the four varieties. After 24 h of storage, there was a continuous drop in the protein contents. At 120 h, the values were 0.29 ± 0.09 g 100 ml^{-1} (MYD), 0.14±0.05 g 100 ml^{-1}(WAT), 0.30 ± 0.08 g 100 ml^{-1} (PB 121+), and 0.22 ± 0.11 g 100 ml^{-1} (PB 113+).

15. Vitamin C

Vitamin C is a water-soluble vitamin, naturally present in some foods, and is easily oxidized and degraded under light, heat, or long storage conditions. Hebbar et al., (2020) reported 19.6 mg 100 $ml^{-1}$of vitamin C in coconut palm sap collected using a coco-sap chiller and 16.3 mg 100 ml^{-1}in palm sap collected by the traditional method. Xia et al., (2011) reported the vitamin C content of fresh coconut palm sap to be 20.4 mg l^{-1}. It reduced slowly on the first day of fermentation and increased slightly to a value of 20.7 mg l^{-1} on 3rdday. This slight increase in vitamin C content could be due to the presence of yeasts which can synthesize vitamin C under fermentation conditions (Bremus et al., 2006). A sharp decrease in the vitamin C content after 3 days might be due to the reduced activity of this yeast. Vitamin C content remained constant after 5 days, which could be explained by the presence of antioxidant phenolic compounds.

16. Volatiles

Borse et al., (2007) isolated volatiles from fresh and fermented coconut sap by solvent extraction and subsequent GC MS analysis revealed the presence of twelve compounds in fermented coconut sap which comprised more than 95% of the volatiles. Of these, only eight were found in fresh coconut sap. The important aroma-contributing compounds were ethyl lactate, phenyl ethyl alcohol, and farnesol, quantities of all three increased in fermented coconut palm sap. The remaining five compounds were decanoic acid, dodecanoic acid, tetradecanoic acid, palmitoleic acid and palmitic acids. Except for palmitic acid, all the other acids increased several-fold in fermented sap. Ethyl esters of fatty acids, along with isoamyl alcohol were present only in fermented coconut palm sap, and this could have been produced during the fermentation.

Purnomo (2007) isolated five volatile compounds from fresh coconut sap, including 2-butanol (60.26-68.37%), acetic acid (25.83-30.43%), 2-methylcyclohexane (0.66-6.39%), cyclohexyloctane (1.81-4.23%), and 1,4-dimethyl-6,1-butyl acetate (0.91-1.11%). Rakphon et al., (2017) identified 21 compounds in fresh coconut sap, with palmitic acid (2024 $\mu g l^{-1}$), palmitoleic acid (1042 $\mu g\ l^{-1}$), and ethyl lactate (560 $\mu g\ l^{-1}$) being the most predominant. In clarified coconut sap, 13 compounds were identified, with palmitic acid

(342 μg l^{-1}), ethyl lactate (300 μg l^{-1}), and phenyl ethyl alcohol (195 μg l^{-1}) as the major constituents. For fermented coconut sap, 11 compounds were identified, among which palmitoleic acid (14,603 μg l^{-1}), isoamyl alcohol (7467 μg l^{-1}), ethyl lactate (4636 μg l^{-1}), phenyl ethyl alcohol (4189 μg l^{-1}), palmitic acid (2421 μg l^{-1}), and dodecanoic acid (1084 μg l^{-1}) were the most significant.

17. Microbial Changes

Various types of microorganisms are supported by palm sap owing to its sugar content. A total of 166 isolates of yeasts and 39 isolates of bacteria were identified in coconut palm sap by Atputarajah (1986). The largest number of isolates (72%) belonged to genera *Candida, Pichia*, and *Saccharomyces. Saccharomyces chevalieri* accounted for 35% of yeast isolates. Predominant genera were *Bacillus* along with *Enterobacter, Leuconostoc, Micrococcus*, and *Lactobacillus.*

From the coconut, palmyra, and nipa palm saps, 204 yeast isolates were isolated and identified as 15 species in the phylum Ascomycota and a species (one strain) in Basidiomycota. *Saccharomyces cerevisiae, Hanseniaspora guilliermondii*, and *Lachancea thermotolerans* were found in the saps of all three palm species. *Candida tropicalis* and *Pichia kudriavzevii* were obtained from the coconut and palmyra palm saps, *Hanseniaspora vineae, Lachancea fermentati*, and *Pichia manshurica* were present in the coconut and nipa palm saps, whereas *Torulaspora delbrueckii* was found in the palmyra and nipa palm saps. The species with the highest occurrence in the saps of coconut, palmyra, and nipa palms was *S. cerevisiae* with 76.67%, 86.70%, and 100% frequency of occurrence, respectively (Limtong et al., 2020).

According to Okafor (1975), the domination of LAB (*L. plantarum*) formed important components of the bacterial populations, while *L. mesenteries* acidified the sap after the first day of tapping. On the 3rd day, after the increase in alcohol concentration, the population of Acetobacter and Gluconobacter species of acetic acid bacteria increased. This increase was associated with an increase in acetic acid concentration from 0.42 to 0.48% between the 3rd and 4th day and then increased to an acceptable level of 0.6%.

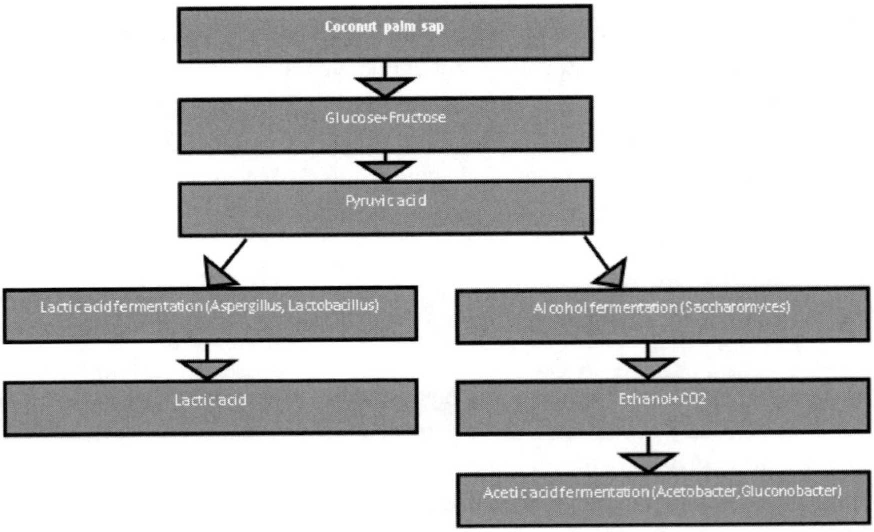

Figure 3. Flow chart showing fermentation reactions in palm sap.

18. Role of Enzymes Present in the Palm Sap

During ambient storage and fermentation, the invertase enzyme present in the palm sap breaks down sucrose into glucose and fructose. These monosaccharides are converted into pyruvate by glycolysis. This process is accompanied by the conversion of two ADP molecules into two ATP and water molecules due to the difference in size of two molecules, namely NAD+ and NADH. With the help of the enzyme pyruvate decarboxylase and cofactor thiamine diphosphate, pyruvate is then converted to acetaldehyde and CO_2. The pyruvate formed by glycolysis is converted to lactic acid by the enzyme lactic acid dehydrogenase. Acetaldehyde is converted into ethanol by the alcohol dehydrogenase (ADH) enzyme (Dioha et al., 2009). The whole conversion of monosaccharide into ethanol through a complex reaction is catalyzed by a group of enzymes called zymase. This sequence of reactions is explained through Figure 3.

Yeasts, particularly *Saccharomyces cerevisiae*, become dominant immediately after tapping. Their growth under ambient conditions can significantly increase the alcohol content within a few hours, contrary to previous reports suggesting a noticeable increase only after three days of fermentation (Amoa-Awua et al., 2007). Research into natural fermentation identified 166 yeast isolates and 39 bacterial isolates, with 17 yeast species from eight genera influencing sap quality during fermentation (Atputharajah et al., 1986).

High-throughput sequencing comparing microbial diversity in fresh *Kalparasa* (pH 7.2) versus fermented sap (pH 4.5) revealed that the former had a higher abundance of bacterial species and lower fungal/yeast diversity than the fermented sap (Gopal et al., 2021). Fresh *Kalparasa* was rich in probiotic-type bacteria like Leuconostoc, and yeasts from the genera Saccharomyces and Hanseniaspora. This suggests that *Kalparasa* fermentation is driven by a symbiotic culture of bacteria and yeasts, including acetic acid bacteria and non-Saccharomyces yeasts. Function predictions of the bacteriome showed enrichment in glycerophospholipid metabolism, ABC transporters, and purine and pyrimidine metabolism, indicating that unfermented sap could be marketed as a healthy, plant-derived food with live probiotic microbiota. In summary, at ambient temperatures (25 ± 5°C), the sap is highly susceptible to microbial and enzymatic fermentation. Refrigeration (6 to 8°C) can delay spoilage for a few hours, while deep freezing (below -20°C) can halt degradation for several months. For extended shelf life, sap should be stored in a deep freezer and transported under refrigerated conditions. It can then be thawed and processed into beverages or other products like coconut sugar.

18. Fermentation Inhibitors

To collect sap, a healthy, mature, unopened inflorescence is selected, and swelling at the base of the inflorescence indicates the optimal collection stage. The inflorescence is then placed into a collection vessel (Borse et al., 2007). As the sap gradually drips out, it is exposed to environmental microorganisms in the open container, leading to spontaneous fermentation. This auto-fermentation transforms the sap into an alcoholic beverage known as toddy (Hebbar et al., 2018).

Natural sap is a habitat for indigenous microorganisms such as *Bacillus*, *Lactobacilli*, *Micrococci*, *Enterobacter*, *Leuconostoc*, *Saccharomyces*, *Candida*, and *Pichia* (Atputharajah et al., 1986). These microbes rapidly ferment the sap, affecting its palatability by producing astringency, unpleasant volatiles, and a cloudy appearance, which can lead to economic losses for farmers (Aparajhitha and Mahendran 2019; Borse et al., 2007). To prevent fermentation in the phloem sap of palm trees, various substances are used to produce economically important products such as sugar, maple syrup, jaggery, and a range of alcohol-free soft drinks. Different methods are employed to inhibit undesirable microorganisms, utilizing both natural and synthetic preservatives, each having its unique advantages and disadvantages (Adisetya and Krisdiarto 2022; Hebbar et al., 2015; Raharjo et al., 2023; Saidan et al., 2020; Singsoong et al., 2010; Somawiharja et al., 2018).

18.1. Plant Materials

Several natural preservatives can serve as alternatives to extend the shelf life of coconut sap, including betel leaves, clove leaves, guava leaves, secang wood (*Caesalpinia sappan* L.), and tea leaves. These materials contain bioactive components that exhibit antimicrobial activity. The use of these preservatives has a positive impact during the tapping process, as they help prevent damage to the coconut sap and mitigate the negative effects of high levels of ash, insoluble materials, and reducing sugars in palm sugar, which can compromise product quality (Haryanti and Mustaufik 2011).

Various bark extracts, such as those from *Saccoglottis gabonensis*, *Vernonia amygdalina*, *Euphorbia* spp., *Nauclea* spp., and *Rubiaceae* spp., were also tested but with limited success (Ogbulie et al., 2007). The stem bark of *Hopea beccariana* Burck is utilized in coconut sugar processing in West Kalimantan, Indonesia, while in other regions, different species such as *Sacoglottis gabonensis* and *Alstonia boonei* in Ikot Ekpene, Akwa Ibom State, Nigeria (Elijah et al., 2010), *Payorm* (*Shorea roxburghii* G. Don) in Thailand (Singsoong et al., 2010), and *Kiam wood* (*Cotylelobium lanceotatum* Carih) also in Thailand (Naknean et al., 2010) are employed.

Haryanti et al., (2012) demonstrated that the optimal combination of laru from green betel leaves (*Piper betle*) and lime resulted in the best physical and chemical properties of coconut sugar when used at 0.75:9.25 ratio, with a natural additive amount of 1 g l^{-1} of coconut sap. The use of 2% calcium hydroxide extended the shelf life of coconut sap by up to 4 h, and mangosteen pericarp proved to be the most effective natural agent for preserving the quality of coconut sap during storage (Naufalin et al., 2012).

Using clove leaves @ 4.5% resulted in the highest pH value, sucrose levels, sensory evaluation scores, and inhibition of spoilage after storage for 4 to 8 h, with guava leaves and mangosteen rind following closely (Sulistyaningrum et al., 2015). Setyawan and Ninsix (2016) incorporated mangrove bark, guava leaves, powdered tea, and betel lime water as preservatives at a concentration of 0.2 g l^{-1} of coconut sap in coconut sugar production. Use of betel lime water yielded the best results, achieving a pH of 5.6, moisture content of 4.94%, and sucrose level of 66.49%.

Ramaswamy and Ramaswamy (2017) reported that a mixture of 10% aloe vera and 3% strawberry extract could preserve sap for up to 2 months at 5°C, suggesting that palm sap could also serve as a sports drink due to its phytonutrient content. Mustaufik et al., (2021) observed quality deterioration of sap if stored for more than 6 h after morning collection and 8 h after afternoon collection. Adisetya and Krisdiarto (2022) employed latex from yellow mangosteen (*Garcinia dulcis*) for coconut sap preservation, with a 1% concentration allowing storage for up to 6 h, while concentrations of 1.5% or 2% permitted storage for 12 h. Raharjo et al., (2023) found that the bark extract of *H. beccariana* Burck contained the highest total flavonoid content and antioxidant activity when extracted using hot water at 60°C. The extracted material contains phenolic compounds that can help prevent microbial damage to coconut sap, inhibiting the growth of *Saccharomyces cerevisiae, Lactobacillus plantarum* and *Lactobacillus mesenteroides*, making it a potential natural preservative for coconut sap. Various amounts of mangosteen bark powder were added to palm sap to evaluate shelf-life stability (Lantemona et al., 2023).

In Thailand, traditional preservation of sap involves adding wood chips from *Payom* (*Shorea roxburghii* G. Don) to the collection device. However, modern practices have introduced chemical preservatives to prevent sap deterioration, although inappropriate use of these chemicals poses health risks to consumers. Singsoong et al., (2010) investigated chemical preservatives in coconut sap produced by farmers in Samut Songkhram, Thailand, finding sulfur dioxide in 39% of samples and benzoic acid in 10% of samples, with concentrations ranging from 11.01 to 2106.94 $mgkg^{-1}$ and 31.37 to 1384.45 $mgkg^{-1}$, respectively. These findings underscore the need to promote natural preservatives to enhance the safety of products made from coconut sap.

Kapilan et al., (2015) incorporated hal bark (*Vateria copallifera*), kahata bark (*Careya arborea*), leaves of neem (*Azadirachta indica*), and lime into collecting pans before use. Lime usage resulted in the lowest alcohol content in coconut sap (which maintained an alkaline pH) as well as reduced counts of individual bacteria and yeast. The pH decreased with the use of hal, kahata, and kohomba bark compared to the control, yielding high-quality sweet sap with lime at 3 g l^{-1}.

In Indonesia, coconut farmers commonly use mangosteen peel and jackfruit wood as preservatives during tapping. The extract of mangosteen is rich in isoprenylated xanthones, which exhibit anticancer, analgesic, neuroprotective, antidiabetic, and hypolipidemic effects (Messi et al., 2014). Tjahjania et al., (2014) observed the potential antioxidant properties of mangosteen rind products. Mangosteen peel contains high levels of polyphenolic compounds (3.717 μgml^{-1}) known as xanthonoids like α-mangostin and γ-mangostin, as well as flavonoids (2.98 μgml^{-1}) (Nivetha and Roy 2015). Xanthones can be isolated from the peel, whole fruit, bark, and leaves of the mangosteen, and various studies have demonstrated their remarkable biological activities, including antioxidant, anti-tumoral, anti-inflammatory, antiallergy, antibacterial, antifungal, and antiviral properties (Chaverri et al., 2008; Palakawong et al., 2010; Suksamrarn et al., 2006).

Haryanti et al., (2018) determined that the optimal preservative concentration of mangosteen peel powder was 0.84 g l^{-1} of coconut sap harvested in sunny weather conditions. The addition of mangosteen peel powder resulted in coconut sap with lower reducing sugar content, higher sucrose levels, and increased antioxidant activity. The total free amino acids (FAA) also decreased with increased mangosteen peel powder, yielding sap with the chemical properties of pH 6.7, TSS 16.2 °Brix, water content 85.62%, reducing sugar 0.04 g 100 g^{-1}, sucrose content 13.64 g 100 g^{-1}, total FAA 0.1738 g 100 g^{-1}, total phenolic content 323.23 mg GAE100 g^{-1}, and vitamin C 26.06 mg 100 g^{-1}.

Saidan et al., (2020) evaluated properties like pH, total acidity, TSS, color, volatile organic compounds (VOCs), and antimicrobial properties of coconut sap with and without the presence of chengal wood chips for ten weeks at 4°C and 25°C storage temperatures. Coconut sap with chengal wood chips kept at 4°C had the lowest total plate count from 1st week until 7th week, while the total plate count of coconut sap without chengal wood chips was too numerous to count throughout the weeks. Chengal wood chip extracts inhibited the growth of *Staphylococcus aureus*, *Escherichia coli*, *Candida albicans,* and *Aspergillus brasiliensis*. The addition of chengal wood to coconut sap slowed down the fermentation process and inhibited the growth of microorganisms.

Haryanti et al., (2023) determined that the optimal concentration of mangosteen peel powder for preserving coconut sap was 0.84 g l-1, specifically for sap tapped during the daytime. This resulted in sap with (all on a dry basis) ash content: 3.51%; protein content: 2.69%; total lipid: 0.10%; total carbohydrate content: 94.62%; water content: 84.21%. In contrast, sap collected during night time contained higher levels of glucose and fructose, measuring 0.85 g and 1.04 g 100 g^{-1}, respectively, compared to lower values found without any treatment: 0.52 g and 0.58 g 100 g^{-1}. Kluz et al., (2023) characterized the mangosteen fruit, noting its high polyphenol content, with values of 3.22 ± 0.68 mg $GAEg^{-1}$ for fresh fruit and 2.17 ± 0.64 mg $GAEg^{-1}$ for the powdered form. Additionally, the antioxidant capacities of the fruit were measured at 21.18% for the fresh fruit and 14.46% for the freeze-dried fruit. These findings highlight the potential of mangosteen peel powder not only in preserving coconut sap but also in enhancing its nutritional and antioxidant properties.

18.2. Chemicals

Baliga and Ivy (1961) found that sodium benzoate, along with citric and malic acids, was effective in preserving coconut sap; however, the addition of malic acid resulted in an off-flavor similar to apple juice. Adding 0.05% sodium metabisulfite followed by pasteurization reduced bacterial populations, though metabisulfite alone was ineffective due to high innate microflora and excessive amounts needed (Faparusi 1969). E-class food-grade preservatives like sodium metabisulfite, calcium hydroxide, and sodium benzoate were shown to control fermentation effectively at ambient temperatures after 48 h of storage (Hariharan et al., 2014).

Okafor (1977) suggested that 1% sorbic acid combined with pasteurization could help preserve palm sap, and pasteurization at 70°C for 30 min without chemical preservatives effectively reduced microbial load. Various clarifying agents, such as gelatine, chitosan, bentonite, and polyvinyl poly pyrrolidone (PVPP), combined with vacuum evaporation and open pan heating, were effective in concentrating the sap. Chitosan, in particular, increased DPPH scavenging activity and was reported as an effective preservation method when used with pasteurization (Naknean et al., 2015).

The use of preservatives is critical in the production of coconut sap to control microbial activity and enhance ethanol production. Samarajeewa et al., (1985) found that adding up to 200 mgl^{-1} of sodium metabisulfite to coconut sap effectively suppressed non-ethanol-producing microorganisms, thereby allowing pure yeast cultures to produce ethanol more efficiently. They recommended using sodium metabisulfite in sap collection pots, either in solution form or as convenient pellets, to mitigate microbial activity and improve ethanol yields. To manage rapid fermentation, Hariharan et al., (2014) suggested adding "E-class" preservatives, including sodium metabisulfite, sodium benzoate, and calcium hydroxide, at various levels to sap collection pots attached to the coconut trees before harvesting. Calcium hydroxide (lime milk) alkalizes the sap by providing hydroxyl ions, which inhibit microorganism growth. The antimicrobial efficacy of lime is influenced by how quickly it dissociates into calcium and hydroxyl ions; the latter raises the pH level, effectively destroying or inactivating various microorganisms (Silveira et al., 2011).

Additionally, Haryanti et al., (2018) explored a natural preservative composed of a mixture of powdered green betel leaves and stems with lime powder. The optimal combination was found to be 2.5% green betel powder and 97.5% lime powder, applied at a rate of 1 g l^{-1} of coconut sap. This treatment produced coconut sugar with favorable characteristics, including reduced sugar content: 12.72% (based on dry basis); water content: 9.45% (based on dry basis); texture: 0.0021 mm.detg^{-1}; ash content: 2.46% (based on dry basis; 2.23% on wet basis). These findings highlight the effectiveness of both synthetic and natural preservatives in enhancing the quality and safety of coconut sap and its derived products. Lime (3 g l^{-1}) was particularly effective in inhibiting microbial growth, thus increasing the sap's shelf life (Ramalakshmi et al., 2018).

The preservation of coconut sap is crucial for maintaining its quality and extending its shelf life, as evidenced by the findings of Somawiharja et al., (2018) in East Java, Indonesia. Their research highlighted that the traditional methods of tapping, collecting, and processing fresh coconut sap significantly impact the quality of both the sap and the resulting sugar. In samples of fresh sap without preservatives, the sugar contents were: sucrose: 1.76%; fructose: 5.76%; glucose: 4.46%. In contrast, sap collected using limestone powder (10 gl-^{1}) showed improved sugar contents such as sucrose: 5.76%; fructose: 3.23%; glucose: 2.25%. These results indicate that fresh coconut sap without preservatives is prone to fermentation, leading to a decrease in sucrose and an increase in fructose and glucose. This trend was similarly observed in coconut sugar made from unpreserved sap. Several preservation methods can enhance the shelf life of coconut sap. One effective approach involves the use of nisin, a natural antimicrobial peptide that protects sap from fermentation. Naknean (2013) reported that combining nisin with calcium carbonate (3000 ppm and 50 ppm, respectively) effectively preserved the physicochemical and sensory qualities of pasteurized coconut sap for up to 21 days when stored at 4°C. This combination resulted in significant reductions in microbial counts [total bacteria: 120×10^5 CFUml^{-1} and total yeast: 143×10^2 CFUml^{-1}]. The treatment also produced favorable physicochemical parameters such as pH: 10.45 ± 0.05; TSS: 15.43 ± 0.12°Brix; total acidity: 1.11 ± 0.04 mg l^{-1}. These promising results suggest that the use of nisin and calcium carbonate could facilitate the industrial-scale preservation of coconut sap, improving its safety and quality for consumer use. An anti-fermenting solution (AFS) combining citric acid (5 mM) and potassium metabisulfite (2 mM) was developed, preserving the sap's nutrients and extending its shelf life to 6 months at ambient temperature (Chinnamma et al., 2019).

18.3. Cold Plasma Bubbling

The preservation of coconut sap is essential for maintaining its quality and safety, and various technologies have been explored to enhance its shelf life. According to Chinnamma et al., (2019), natural sap can be effectively preserved through a combination of preservatives such as citric acid and sodium metabisulphite, along with processing methods like centrifugation, filtration, and pasteurization. Additionally, a patented preservation technology involves acidification and the use of preservatives like nisin combined with pasteurization, as noted by Ramalakshmi et al., (2004).

While these technologies significantly improve the shelf life of coconut sap, they may also compromise various quality attributes due to thermal pasteurization. For instance, Chia et al., (2012) highlighted that thermal treatments could negatively impact the sensory qualities and nutritional content of the sap, potentially decreasing consumer acceptance.

To address these challenges, alternative preservation methods, such as cold plasma treatment, are being investigated. Cold plasma is a non-thermal processing technique that can effectively inactivate microbes at room temperature, thus minimizing negative effects on color, nutrition, and other food properties (Dasan and Boyaci 2018). During the plasma bubbling process, a stable and transient bubble is formed, leading to the dissociation of water (approximately 80%) into hydroxyl radicals (OH) and hydrogen atoms (H_2). These reactive species primarily exist within the bubble, similar to the behavior of radicals in an ultrasound field (Margulis 1976).

Aparajhitha and Mahendran (2019) conducted experiments by bubbling cold plasma gas through coconut sap at various input voltages and flow rates to assess changes in microbial and physicochemical properties. Their findings revealed a considerable increase in pH (~1.04 increase), negligible changes in color and TSS values, however the nutritional profile remained largely unchanged, with only a slight decrease in vitamin C content (~4.7 mg 100 ml^{-1}). Also, a significant reduction in microbial count was observed, indicating effective microbial inactivation through cold plasma treatment. This innovative approach holds promise for preserving coconut sap while maintaining its desirable qualities, potentially improving consumer acceptance and extending shelf life without compromising nutrition.

18.4. Frozen Storage

Frozen storage is highly effective in preserving the quality and extending the shelf life of *neera*. At temperatures around -20°C, the total viable microbial count, including yeast and LAB, is significantly reduced. Specifically, frozen storage at -20°C results in a reduction of 6.26 log CFU ml^{-1} for the total viable count and 3 log CFU ml^{-1} for yeast, while inhibiting the growth of lactic acid bacteria which are crucial in fermentation. This storage condition also minimizes chemical changes, including pH (5.37), TSS (14.0°Brix), and total acidity (0.82 mg l^{-1}), while maintaining color, viscosity, total protein, and ethanol content close to those of fresh *neera*. Live/dead cell fluorescence staining further supports that frozen storage at -20°C effectively distinguishes between injured and resistant microbes, confirming its role in preserving *neera*'s quality. Sensory attributes remain acceptable for up to 21 days under these conditions. Thus, frozen storage at -20°C provides a practical solution for maintaining *neera's* freshness and safety over extended periods, making it fit for consumption and potentially beneficial for commercial and domestic use (Sukumaran and Radhakrishnan 2021).

18.5. Thermal Processing

Thermal processing can affect the sap's nutritional content, color, and flavor due to Maillard reactions, inversion, and caramelization (Ho et al., 2007). Pasteurization at 76-80°C did not alter the natural flavor, whereas higher temperatures (82-88°C for 4-5 min) imparted a cooked flavor (Baliga and Ivy 1961). Yeasts are destroyed by pasteurization of 25-30 min. The presence of air increases the fermentation. Hebbar et al., (2018) reported that pasteurizing at 90-95°C for 3-7 minutes or boiling at 60-70°C for 60 minutes could preserve palm sap. Mohanadas (1974) found that sterilizing and bottling the sap at 80°C and 90°C for 25 and 20 min, respectively, was effective in preserving it for up to 6 months without altering the flavor.

Proper handling during tapping, transportation, and storage is crucial before applying preservation techniques. (Pandiselvam et al., (2020) studied microwave treatment (600 and 900 W) of coconut inflorescence sap, finding that 900 W for 120 seconds maintained the sap's physicochemical properties for up to 16 days. Pulsed light (PL) treatment for the preservation of coconut inflorescence sap reduced the microbial population, maintained quality parameters like pH, TSS, color, and turbidity, and preserved the nutritional profile of the coconut palm sap (Priya et al., 2024).

In order to stop the fermentation of the sap and maintain the innate quality, the microbial growth should be controlled. Among the different preservation methods, microwave heat treatment has advantages of volumetric heating and higher penetration depth. Rajan et al., (2019) attempted microwave treatment of coconut sap. The pH of fresh sample was 6.09 and the values decreased with different power levels and exposure time. However, the changes were less with all power levels (520 W, 720 W and 900 W) with the exposure time of 3 min. being 5.57, 5.73 and 5.70, respectively. The acidity values increased with increasing the power levels and exposure times. The acidity of fresh sap was 0.38 (g citric acid l^{-1}) and the highest acidity value (0.64) was found with the exposure time of 9 min with power levels of 720 or 900 W. The TSS of the coconut sap increased with increasing the power levels and time periods, the values for exposure time of 9 min for the three power levels being 24.10, 31.93 and 42.20°B respectively. The fresh sap had TSS of 16.87°B. Both the reducing sugar and total phenolic contents, from the initial values of 12.02 mg glucose ml^{-1} and 0.202 mg gallic acid ml^{-1} respectively, increased to 20.26 mg glucose ml^{-1} and 0.620 mg gallic acid ml^{-1}, respectively with 900 W power level and 9 min exposure time. Coconut sap is a heat sensitive product, and hence, controlling the temperature and treatment time or high temperature short time process are necessary to retain various physico chemical parameters.

The study by Somashekaraiah et al., (2019) highlights the promising use of probiotic strains isolated from *neera* for potential applications in food and feed formulations. The key findings include; seven probiotic bacterial isolates were screened from *neera*, all demonstrating resistance to gastrointestinal conditions and significant antimicrobial activity. Among these isolates, *Lactobacillus brevis* MYSN 106 was identified as the most effective. It exhibited robust survival in the acidic and bile-rich environments of the stomach, as well as the challenging conditions of the intestines. *L. brevis* MYSN 106 also showed surface-binding capabilities, which are essential for effective colonization of the gastrointestinal tract. This colonization ability is crucial for its antimicrobial action and potential therapeutic benefits. The Lactobacillus isolates from *neera* possess significant probiotic attributes and antimicrobial activities, suggesting their potential use as probiotics in various food and feed products. In summary, *neera*-derived probiotic strains, particularly *L. brevis* MYSN 106, offer substantial promise for enhancing food safety and health benefits through their antimicrobial and gastrointestinal resilience.

Concluding Remarks and Prospects

Palm sap quality varies with location and tapping duration, presenting preservation challenges such as oyster white coloration, harsh flavor, and taste due to its fermentable nature. This chapter explores the principal biochemical reactions and microbiome involved in the fermentation process of coconut and palm inflorescence sap. Also, various techniques to enhance the quality and safety of palm sap while preserving its natural flavor are also discussed. The primary objective of fermentation chemistry is to identify microbes responsible for both desirable and undesirable flavors. This scientific understanding is crucial for ensuring the quality and safety of palm juice through technological interventions. Consequently, methods to extend the sap's shelf life have been developed. One of the significant advancements is the coco-sap chiller technology developed by ICAR-Central Plantation Crops Research Institute (CPCRI), which has addressed the challenge of collecting hygienic, unfermented sap. The collected sap is highly nutritious and can be sold fresh under refrigeration or processed into products like coconut sugar, jaggery, nectar, or syrup without the use of chemicals, preservatives, or additives. Future research directions could focus on studying the changes in the flavor profile of *Kalparasa* during honey and sugar production. Additionally, comparing the quality profiles of honey and sugar made from *Kalparasa* using different processing methods, such as open evaporation and vacuum evaporation, may prove beneficial for producing high-quality jaggery products. Food industries must meet consumer demand for convenient, fresh, and healthy products without synthetic preservatives. To achieve this, non-thermal processing methods such as ozonation, high-pressure processing, pulsed electric fields, cold plasma techniques, UV light treatment, and pulsed light can be used to preserve palm sap. Employing non-thermal techniques and minimizing chemical preservatives for microbial control in palm sap could expand its accessibility to customers in non-traditional markets and help realize its full commercial potential.

References

Adisetya, E., Krisdiarto, A.W. (2022) Preservative of coconut sap shelf life derived from mangosteen yellow latex. *JurnalIlmiahTeknologi Dan Industri Pangan Unisri* 7(1)" 59–67. https://doi.org/10.33061/jitipari.v7i1.6757.

Aguilera, J., Gil, R.F., Prieto, J.A. (2007) Cold response in Saccharomyces cerevisiae: new functions for old mechanisms. *FEMS Microbiol Rev* 31:327–341, doi:10.1111/j.1574-6976.2007.00066.x.

Amoa-Awua, W. K., Sampson, E., Tano-Debrah, K. (2007). Growth of yeasts, lactic and acetic acid bacteria in palm wine during tapping and fermentation from felled oil palm (Elaeis guineensis) in Ghana. *J Applied Microbio*l 102 (2): 599–606. doi:10.1111/j.1365-2672.2006.03074.x.

Aparajhitha, S., Mahendran, R. (2019). Effect of plasma bubbling on free radical production and its subsequent effect on the microbial and physicochemical properties of coconut neera. Innovative *Food Sci Emerging Technol* 58, 102230. https://doi.org/10.1016/j.ifset.2019.102230.

Atputharajah, J.D., Widanapathirana, S., Samarajeewa, U. (1986). Microbiology and biochemistry of natural fermentation of coconut palm sap. *Food Microbiol* 3 (4): 273–280. https://doi.org/10.1016/0740-0020(86)90009-2.

Baliga, B.P, Ivy, A.C. (1961). Beverage preservation, pasteurization of palm sap (*Neera*). *J Agric Food Chem* 9 (2): 149–151.https://doi.org/10.1021/jf60114a018.

Bashyam, R., Kulathooran, R., Nagarajarao, R. M., Borse, B. B. Vishweshwaraiah P (2004) A process for the preservation of deodourised coconut sap (*Neera*). *Indian patent* - WO2004054389 A1.

Borse, B.B., Rao, L.J.M., Ramalakshmi, K., Raghavan, B. (2007). Chemical composition of volatiles from coconut sap (neera) and effect of processing. *Food Chem* 101(3): 877–880. doi:10.1016/j.foodchem.2006.02.026.

Bremus, C., Herrmann, U., Bringer-Meyer, S, Sahm, H. (2006). The use of microorganisms in L-ascorbic acid production. *J Biotechnol* 124 (1): 196–205. doi:10.1016/j.jbiotec.2006.01.010.

Chaverri, J.P, Rodriguez, CN, OrozcoIbarra, M., Rojas, P.J.M. (2008). Medicinal properties of mangosteen (*Garcinia mangostana*). *Food Chem Toxicol* 46(10): 3227-3239, doi: 10.1016/j.fct.2008.07.024.

Chia, S.L., Rosnah, S., Noranizan, M.A., Ramli, W.W.D. (2012). The effect of storage on the quality attributes of ultraviolet irradiated and thermally pasteurised pineapple juices. *Int Food Res J.*19(3):1001-1010

Chinnamma, M, Bhasker, S., Binitha Hari, M., Sreekumar, D., Madhav, H. (2019). Coconut neera—a vital health beverage from coconut palms: harvesting, processing and quality analysis. *Beverages* 5 (1): 22. doi:10.3390/beverages5010022.

Damiani, C., Lage, M.E., Silva, F.A.D., Pereira, D.E.P., Becker, F.S., Boas, E.V.B.V. (2013) Changes in the physicochemical and microbiological properties of frozen araca pulp during storage. *Food Sci Technol* 33(1): 19–27.

Dasan, B.G., Boyaci, I.H. (2018). Effect of cold atmospheric plasma on inactivation of *Escherichia coli* and physicochemical properties of apple, orange, tomato juices, and sour cherry nectar. *Food and Bioprocess Technol* 11(2): 334–343. https://doi.org/10.1007/s11947-017-2014-0.

Dioha, I.J., Olugbemi, O., Odin, E., Eneji, M.A. (2009). Zero additives preservation of Raphia palm wine. *Int J Biol Chem Sci* 3 (6). Available from: 10.4314/ijbcs.v3i6.53144.

Elijah, A.I., Ojimelukwe P.C., Ekong U.S., Asamudo N.U. (2010). Effect of *Sacoglottis gabonensis* and *Alstonia boonei* on the kinetics of *Saccharomyces cerevisiae* isolated from palm wine. *Afr J Biotechnol* 9(35):5730–5734.

Eze, M.O., Ogan, A.U. (1988). Sugars of the unfermented sap and the wine from the oil palm, *Elaeis guinensis*, tree. *Plant Foods for Human Nutrn* 38 (2):121–126.

Faparusi, S.I. (1969). Effect of pH on the preservation of palm wine by sulfite. A*pplied Microbiol* 18 (1):122–123. doi.org/10.1128/am.18.1.122-123.1969.

Faparusi, S.I. (1974). Microorganisms from oil palm tree (Elaeis guineensis) tap holes. J*ournal of Food Science* 39 (4):755–757, doi.org/10.1111/j.1365-2621.1974.tb17972.x.

Febriyanti, R., Susanto, W. H… and Nugrahini N. P. (2015) Karakteristik sirup jahe nira kelapa terfermentasi delapan jam (kajian jenis dan konsentrasi sari jahe). *J. Pangan dan Agroindustri* 3(3):1026–1031.

Ghosh D. K., Bandyopadhyay A., Das S., Hebbar K.B., Biswas B. (2018). Coconut sap (Neera)-untapped opportunity of spinoff gains in West Bengal, India. *Int J Current Microbiol Applied Sc*i 7 (9): 1883–1897.

Gopal, M., Shil, S., Gupta, A., Hebbar, K.B., Arivalagan, M. (2021). Metagenomic investigation uncovers presence of probiotic-type microbiome in Kalparasa®(fresh unfermented coconut inflorescence sap). *Frontiers in Microbiol* 12, 662783. doi:10.3389/fmicb.2021.662783.

Hariharan, B., Singaravadivel, K., Alagusundaram, K. (2014). Effect of food grade preservatives on the physicochemical and microbiological properties of coconut toddy during fermentation. *J Nutrition Food Sci* 4(5): 1–5. doi:10.4172/2155-9600.1000299.

Haryanti, P., and Mustaufik, (2011) Evaluasi mutu gula kelapa kristal (gula semut) di kawasan home industri gula kelapa kabupaten Banyumas (Quality Evaluation of Coconut Crystal Sugar at Household Industry in Banyumas Regency) *Jornal Agroteknologi* 5: 48–61.

Haryanti, P., Karseno, K., Handayani, I., Sulistyo, S.B. (2023)The chemical composition of coconut sap at different tapping condition. AIP Conf. Proc. 2586, https://doi.org/10.1063/5.0107040.

Haryanti, P., Karseno, K., Setyawati, R. (2012). Application of natural preservatives for coconut made from green betel to physical and chemical properties of coconut sugar. *J Pembang Pedesaan* 12:106-12.

Haryanti, P., Supriyadi, Marseno D.W., Santoso, U(2018) Effects of different weather conditions and addition of mangosteen peel powder on chemical properties and antioxidant activity of coconut sap. *Agritech* 38 (3):295-303, https://doi.org/10.22146/agritech.29844.

Hebbar, K., Arivalagan, M., Manikantan, M., Mathew, A., Thamban, C., Thomas, G.V., Chowdappa, P. (2015) Coconut inflorescence sap and its value addition as sugar–

collection techniques, yield, properties and market perspective. *Current Sci*109(8):1411-1417.

Hebbar, K.B., Arivalagan, M., Pavithra, K.C., Roy, T.K., Gopal, M., Shivashankara, K.S., Chowdappa, P. (2020). Nutritional profiling of coconut (*Cocos nucifera* L.) inflorescence sap collected using novel coco-sap chiller method and its value added products. *J Food Measurement and Characterization* 14:2703–2712. doi:10.1007/s 11694-020-00516-y.

Hebbar, K.B., Mathew, A.C., Arivalagan, M., Samsudeen, K., Thomas, G.V. (2013). Value added products from Neera. *Indian Coconut J.* 56(4): 28–33. doi:10.1007/978-3-662-04857-3.

Hebbar, K.B., Pandiselvam, R., Manikantan, M.R., Arivalagan, M., Beegum, S., Chowdappa, P. (2018). Palm sap—quality profiles, fermentation chemistry, and preservation methods. *Sugar Tech* 20 (6): 621–634. doi:10.1007/s12355-018-0597-z.

Ho, C.W., Aida, W.M.W., Maskat, M.Y., Osman, H. (2007). Changes in volatile compounds of palm sap (Arenga pinnata) during the heating process for production of palm sugar. *Food Chem* 102 (4): 1156–1162. https://doi.org/10.1016/j.foodchem. 2006.07.004.

Iwuoha, C.I., Eke, O.S. (1996). Nigerian indigenous fermented foods: their traditional process operation, inherent problems, improvements and current status. *Food Res Inter.* 29(5-6):527-40.

Jackson, J.F., Linskens, H.F. (2002). Analysis of taste and aroma [online]. Springer Science & Business Media. Available from: doi:10.1007/978-3-662-04857-3.

Jayashree, A. (2013) Focus on commercialisation of technologies to capitalise the opportunities for neera. *Indian Coconut J* 56(1):27–31.

Kadere, T.T., Oniango, R.K., Kutima, P.M., Muhoho, S.N. (2004). Traditional tapping and distillation methods of coconut wine (mnazi) as practised in the coastal region of Kenya. *African J Food Agric Nutr Dev* 4:1–16. doi: 10.4314/ajfand. v4i1.19153.

Kalaiyarasi, K., Sangeetha, K., Rajarajan, S. (2013). A comparative study on the microbial flora of the fresh sap from cut inflorescence and fermented sap (toddy) of *Borrassus flabellifer* Linn. (palmyrah tree) and of *Cocos nucifera* Linn. (coconut tree) to identify the microbial fermenters. *Int J Res in Pure and Applied Microbiol* 3(3): 43-47.

Kapilan, R., Kailayalingam, R., Mahilrajan, S., Srivijeindran, S. (2015) Determination of efficient fermentation inhibitor of sweet sap of *Cocos nucifera* and optimization of concentration for quality outputs in northern Sri Lanka. *Int J Curr Microbiol App Sci* 2(7):166-174, doi: 10.12983/ijsras-2015-p0166-0174.

Khadka, N., Acharya, D.R., Dangal, A., Rai, K., Gurung, G., Sherma, G., Khatri, S.B., Gautam, N. (2024). Study on the changes during the fermentation of the wine prepared from palm (*Phoenix sylvestris*) sap. *Heliyon* doi: 10.1016/j.heliyon.2024.e35799.

Kluz, M., Pietrzyk, K., Puchalski, C. (2023) Antioxidant and antimicrobial properties of *Garcinia mangostana. Acta Hort Regiotec* 26: 43–47, doi: 10.2478/ahr-2023-0007.

Kurniawan, T., Jayanudin, Kustiningsih I., Firdaus, M.A. (2018) Palm sap sources, characteristics, and utilization in Indonesia. *J Food and Nutrn Res* 6(9): 590-596, doi:10.12691/jfnr-6-9-8.

Lantemona, H., Sahupala, R., Boka, R.Y. (2023). Mangosteen (*Garcinia mangostana* L.) trunk bark as palm sap preservative. *Food Res* 7 (3): 98–104. doi:10.26656/fr.2017. 7(3).942.

Leena, M.M., Yoha, K.S., Moses, J.A., Anandharamakrishnan, C. (2021) Electrospun nanofibrous membrane for filtration of coconut *neera*. *Nanotechnol Environ Eng* 6:24, https://doi.org/10.1007/s41204-021-00116-1.

Levang, P. (1988) Le cocotier est aussi une plante sucridlre. *Oléagineux* 43(4): 159–164

Limtong, S., Am-In, S., Kaewwichian, R., Kaewkrajay, C., Jindamorakot, S. (2020). Exploration of yeast communities in fresh coconut, palmyra, and nipa palm saps and ethanol-fermenting ability of isolated yeasts. *Antonie Van Leeuwenhoek* 113: 2077–2095. doi:10.1007/s10482-020-01479-2.

Margulis, M. (1976). A. Cavitation-diffusion model of the spatial distribution of radicals in an ultrasonic field. *Russian J Physical Chem* 50: 534–537.

Mat, K., Kari, A.Z., Rusli, N.D., Harun, C.H., Wei, L.S., Rahman, M.M., Khalid, M.H.N., Hanafiah, M.A.M.H., Sukri, M.S.A., Khalif, R.R.I.A., Zin Z.M., Zainol, M.K.M., Panadi, M., Nor, M.F.M., Goh, K.W. (2022) Coconut palm: food, feed, and nutraceutical properties. *Animals* 12(16), 2107, https://doi.org/10.3390/ani12162107.

Messi, B.B., Ho, R., Lannang, A.M., Cressend, D., Perron, K., Nkengfack, A.E, Carrupt, P.A., Hostettmann, K., Cuendet, M. (2014) Isolation and biological activity of compounds from *Garcinia preussii*. *Pharmaceutical Biol* 52(6):706–711, 10.3109/13880209.2013.865241.

Mohanadas, S. (1974). Preservation bottling and keeping qualities of fresh coconut sap (Sweet toddy). Ceylon Cocon. *Quart.* 25:109–115. doi:10.1007/s10482-020-01479-2.

Mustaufik, S. L., Rahayoe, S., Widodo, K.H. (2021). The effect of time and duration of tapping and the addition of laru as natural preservative in coconut sap quality. IOP *Conference Series: Earth and Environmental Science*, 653(1): 012084 https://doi.org/10.1088/1755-1315/653/1/012084.

Mustaufik, S.L, Rahayu, S., Widodo, K.H. (2022) Technique engineering of tapping and shelter of coconut sap and its effect on the quality of crystal coconut sugar. *Food Research* 6 (2) : 248 – 254, doi: https://doi.org/10.26656/fr.2017.6(2).220.

Naik, J.B., Suresh, P.R., Manjusha, M., Balachandran, P. V., Balakrishnan, P.C. (2013) 'Keraamritham'-a health drink from coconut inflorescence sap. *Indian Coconut J* 56(1):9–10.

Naknean, P. (2013). Improvement in shelf life and safety of pasteurized palm sap (Borassus *flabellifer* L.) by the addition of nisin. *J Food Safety* 33(4):515–525. doi:10.1111/jfs.12084.

Naknean, P., Jutasukosol, K, Mankit, T. (2015). Utilization of chitosan as an antimicrobial agent for pasteurized palm sap (*Borassus flabellifer* Linn.) during storage. *J Food Sci Technol* 52:731–741. doi: 10.1007/s13197-013-1104-x.

Naknean, P., Meenune, M., Roudaut, G. (2010). Characterization of palm sap harvested in Songkhla province, Southern Thailand. *Int Food Res J* 17(4):977–986.

Naufalin, R., Yanto, T. and Binardjo, A.G. 2012Penambahan konsentrasi Ca(OH)2 dan bahan pengawet alami untuk peningkatan kualitas nira kelapa. *Jurnal Pembangunan Pedesaan* 12(2):86-96.

Nivetha, S., Roy, D.V. (2015). Antioxidant Activity and Antimicrobial Studies on *Garcinia mangostana. Am J Biol Pharmaceutical Res* 2(3): 129–134.

Nwachukwu, I.N., Ibekwe, V.I., Nwabueze, R.N., Anyanwu, B. N. (2006). Characterisation of palm wine yeast isolates for industrial utilisation. *African J Biotechnol* 5 (19):1725–1728.

Ogbulie, T.E., Ogbulie, J.N., Njoku, H.O. (2007). Comparative study on the microbiology and shelf life stability of palm wine from *Elaeis guineensis* and *Raphia hookeri* obtained from Okigwe, Nigeria. *African J Biotechnol* 6 (7): 914–922.

Okafor, N. (1975). Microbiology of Nigerian palm wine with particular reference to bacteria. *J Applied Bacteriol* 38 (2): 81–88. doi:10.1111/j.1365-2672.1975.tb00507.x.

Okafor, N. (1977). The microbiological basis of a method for palm wine preservation. *J Applied Bacteriol* 43 (1): 159–161. doi:10.1111/j.1365-2672.1977.tb00735.x.

Palakawong, C., Sophanodora, P., Pisuchpen, S., Phongpaichit, S. (2010). Antioxidant and antimicrobial activities of crude extracts from mangosteen (*Garcinia mangostana* L.) parts and some essential oils. *In Food Res J* 17(3): 583–589.

Pandiselvam, R., Hebbar, K.B., Manikantan, M.R., Prashanth, B.K., Beegum, P.P.S., Ramesh, S.V. (2020). Microwave treatment of coconut inflorescence sap (Kalparasa®): a panacea to preserve quality attributes. *Sugar Tech* 22:718–726. doi:10.1007/s12355-020-00828-9.

Pandiselvam, R., Manikantan, M.R., Binu, S.M., Ramesh, S.V., Beegum, P.P.S., Gopal, M., Shil, S. (2021). Reaction kinetics of physico-chemical attributes in coconut inflorescence sap during fermentation. *J Food Sci Technol* 58: 3589-3597, https://doi.org/10.1007/s13197-021-05088-3.

Penn, M.D., Galgoci, B., Greer, H. (1983). Identification of AAS genes and their regulatory role in general control of amino acid biosynthesis in yeast. Proceedings of the National Academy of Sciences 80 (9): 2704–2708. doi: 10.1073/pnas.80.9.2704.

Prashanth, B.K., Patil, V.S. (2020) Effect of thermal pasteurization on shelf life and quality of *Kalparasa* (Coconut sap). *J Farm Sci* 33(1):1-4.

Priya, V.V., Balakrishnan. M., Pandiselvam, R., Karthikeyan, S., Rajkumar, P., Manikantan, M.R., Gurusamy, K., Preetha, P. (2024). Effect of pulsed light treatment for the preservation of coconut inflorescence sap. *Sugar Tech* 26(1):106–116 doi:10.1007/s12355-023-01312-w.

Purnomo, H. (2007) Volatile components of coconut fresh sap, sap syrup and coconut sugar. *ASEAN Food J* 14(1): 45–49.

Raharjo, D., Zaman, M.Z., Praseptiangga, D., Yunus, A. (2023) Physicochemical and microbiological characteristics of various stem bark extracts of *Hopea beccariana* Burck potential as natural preservatives of coconut sap. *Open Agriculture* 8: 20220175, https://doi.org/10.1515/opag-2022-0175, 19 pages.

Rajan, A., Anandakumar, S., Gowrisree, V., Loganathan, M. (2019) Effect of microwave treatment on physico-chemical quality of coconut neera. *Int J Pure Appl Biosci* 7(1):463–468. https://doi.org/10.18782/2320-7051.7564.

Rakphon, A., Srijesadaruk, V. (2017) Effect of thermal processing on physical, chemical properties and volatile compounds of coconut (*Cocos nucifera* L.) sugar. In *Proceedings of the 4th International Postgraduate Symposium on Food, Agriculture and Biotechnology* (IPSFAB); Maha Sarakham, Thailand, 2017; pp. 80–91.

Ramalakshmi, K., Nidheesh, T., Chanukya, B.S., Jagan Mohan Rao, L. (2018). Changes in quality profile and flavour components of coconut sap during natural fermentation. *Food Nut Curr Res* 1 (3): 70–76.

Ramalakshmi, K., Ramesh, M., Raghavan, B., Prakash, V. (2004) Process for the preservation of coconut sap (neera). U.S. Patent Application No. 10/396710.

Ramaswamy, P., Ramaswamy, L. (2017). Organoleptic acceptability, selected nutrient content and physicochemical characteristics of coconut neera on storage. *J Food Sci Eng* 7(4): 192–201. doi: 10.17265/2159-5828/2017.04.003.

Saidan, N.H., Kartayasa, M.A.A., Lokman, N.I.A.M., Ab Llah, N.F., Nor M.M., Mokhtar, S.I., Zaman, K. A.K.P, Sidek, N.M., Hamil, M.S.R. (2020). Effect of chengal wood (*Neobalanocarpus heimii*) on physicochemical and antimicrobial properties of coconut sap (*Cocos nucifera*). In: IOP Conference Series: Earth and Environmental Science 12073. doi:10.1088/1755-1315/549/1/012073%0A.

Saini, H.S., Rahim, N.S.A., Yaakop, A.S. (2021) Elucidation of aerobic bacteria diversity in *neera* water from coconut and nipa tree. IOP Conf. Series: Earth and Environmental Science 711, 012016, doi:10.1088/1755-1315/711/1/012016, 11 pages.

Samarajeewa, U., Mathes, D.T., Wijeratne, M.C.P., Warnakula, T. (1985) Effect of sodium metabisulphite on ethanol production in coconut inflorescence sap. *Food Microbiol* 2(1):11-17, doi:10.1016/0740-0020(85)90018-8.

Sánchez, D.G., Ríos, G.F.L. (2002) Manejo de la palma de coco (*Cocos nucifera* L.) en México (Management of the coconut palm (*Cocos nucifera* L.) in Mexico). *Rev Chapingo Ser Cienc For Ambiente* 8(1):39-48.

Sanni, A. (1993). The need for process optimization of African fermented foods and beverages. *Int J Food Microbiol*18(2):85-95.

Setyawan, A., Ninsix, R. (2016). Studi penambahan pengawet alami pada nira terhadap mutu gula kelapa yang dihasilkan. *Jurnal teknologi pertanian*, 5(2). https://doi.org/10.32520/jtp.v5i2.90.

Shetty, S.S., Ramesh, S.V., Arivalagan, M., Roopashree, P.G., Manikantan, M.R., Hebbar, K.B., Kumari, S. (2023) Coconut (*Cocos nucifera* L.) inflorescence sap-derived sugar restores the glucose and lipid homeostasis in streptozotocin-induced diabetic Wistar rat model. *Biomedicine* 43(3):859–866, https://doi.org/10.51248/.v43i3.1453.

Silveira, C.F.D.M., Cunha, R.S., Fontana, C.E., Gomes, B.P.F.D.A., Motta, R.H.L., Bueno C.E.D.S. (2011). Assessment of the antibacterial activity of calcium hydroxide combined with chlorhexidine paste and other intracanal medications against bacterial pathogens. *European J Dentistry* 5(1):1–7, 10.1055/s-0039-1698852.

Singsoong, K., Benjapong, W., Nitithamyong, A., Tavichatwitayakul, R., Karnpanit, W., Moungsrichan, N., Prakobsil, J. (2010). Preservatives used in coconut sap. *Toxicology Letters*, 196S, P309-045, doi:10.1016/j.toxlet.2010.03.1079.

Somashekaraiah, R., Shruthi, B., Deepthi, B.V., Sreenivasa, M.Y. (2019). Probiotic properties of lactic acid bacteria isolated from neera: A naturally fermenting coconut palm nectar. *Frontiers in Microbiol* 10:1–11. https://doi.org/10.3389/fmicb.2019.01382.

Somawiharja, Y., Wonohadidjojo, D.M., Kartikawati, M., Suniati, F.R.T., Purnomo, H. (2018) Indigenous technology of tapping, collecting and processing of coconut (*Cocos nucifera*) sap and its quality in Blitar Regency, East Java, Indonesia. *Food Res* 2(4): 1-6, https://doi.org/10.26656/fr.2017.2(4).075.

Suksamrarn, S., Komutiban, O., Ratananukul, P., Chimnoi, N., Lartpornmatulee, N., Suksamrarn, A. (2006). Cytotoxic prenylated xanthones from the young fruit of *Garcinia mangostana*. *Chemical Pharmaceutical Bulletin* 54(3): 301-305, doi: 10.1248/cpb.54.301.

Sukumaran, L., Radhakrishnan, M. (2021). Effect of frozen storage on the inhibition of microbial population, chemical and sensory characteristics of coconut *neera*. *J Applied Microbiol* 131 (4):1830–1839.

Sulistyaningrum, A., Yanto, T., Naufalin, R. (2015) Perubahan kualitas nira kelapa akibat penambahan pengawet alami (alteration of coconut sap quality due to the addition of natural preservatives). *Jurnal Penelitian Pascapanen Pertanian* 12(3): 137-146, doi: 10.21082/jpasca.v12n3.2015.137-146.

Tjahjania, S., Widowatia, W., Khionga, K., Suhendraa, A., Tjokropranotoa, R. (2014) Antioxidant properties of *Garcinia mangostana* L (mangosteen) rind. *Procedia Chemistry* 13: 198 – 203, doi: 10.1016/j.proche.2014.12.027.

Tomomatsu, A., Itoh, T., Wijaya, C.H., Nasution, Z., Kumendong, J., Matsuyama, A. (1996) Chemical constituent of sugar containing sap and brown sugar from palm in Indonesia. *Japanese J Trop Agric* 40(4): 175–181.

Vidanapathirana, S., Atputharajah, J.D., Samarajeewa, U. (1983). Microbiology of coconut sap fermentation. *Vidyodaya J Arts Sci Lett* 11(1& 2): 35-39.

Wellala, C.K.D., Gunawardhane, M.H.W., Wijeratne, M.C.P., Illeperuma, C.K. (2004). Fermentation of rice using yeast isolated from coconut and palmyrah sap. *Agric Res* 16: 114-120.

Wiboonsirikul, J, Ongkunaruk, P., Poonpan, P. (2024) Determining key factors affecting coconut sap quality after harvesting. *Heliyon* 10, e29002, https://doi.org/10.1016/j.heliyon.2024.e29002, 14 pages.

Wijesinghe, D.G.N.G., Samarajeewa, U. (1988). Screening yeasts from coconut inflorescence sap for continuous alcoholic fermentation. *Food Microbiol* 5:119- 123. https://doi.org/10.1016/0740-0020(88)90009-3.

Xia, Q., Li, R., Zhao, S., Chen, W., Chen, H., Xin, B., Huang, Y., Tang, M. (2011). Chemical composition changes of post-harvest coconut inflorescence sap during natural fermentation. *African J Biotechnol* 10 (66): 14999–15005.

Ysidor, K.N., Konan, K.J., Roger, K.B., Assa, R.R., Joelle, O.D.M., Emmanuel, I.A., Marius, B.G.H. (2015) Changes in physicochemical parameters during storage of the inflorescence sap derived from four coconut (*Cocos nucifera* L.) varieties in Côte d'Ivoire. *Ame J of Exper Agri* 5(4):352-365, doi: 10.9734/AJEA/2015/11035.

Chapter 7

Coconut Palm Sap as a Beverage and Other Value-Added Products

M. R. Manikantan[1,*]
V. Krishnakumar[1]
S. V. Ramesh[1]
J. Anjitha[1]
S. Sudharshana[1]
P. P. S. Beegum[1]
D. K. Ghosh[2]
and K. B. Hebbar[1]
[1]ICAR-Central Plantation Crops Research Institute, Kasaragod, Kerala, India
[2]Bidhan Chandra Krishi Viswavidyalaya, Mohanpur, Nadia, India

Abstract

Neera, the natural sap from mature coconut palms, is rich in essential nutrients, minerals, and vitamins which are crucial for human health. *Neera* is traditionally known for its refreshing qualities as a beverage, and also holds significant potential for value addition in various forms. This chapter explores the diverse applications of *neera* from its use as a health drink to the production of fermented beverages, sweeteners like coconut sugar, and other processed products. It also explains the nutritional benefits, processing methods, and challenges involved in advancing *neera* production while promoting sustainable agricultural

[*] Corresponding Author's Email: manikantan.mr@icar.gov.in

In: Coconut Palm Sap
Editors: K. B. Hebbar, V. Krishnakumar and S. V. Ramesh
ISBN: 979-8-89530-640-6

practices. Through this exploration, the chapter aims to underscore the importance of innovation in the coconut industry, providing insights into how coconut palm sap can be transformed into a range of profitable and health-conscious products. By highlighting the multifaceted applications and benefits of coconut palm sap, this chapter discusses its critical role in enhancing agricultural value chains and promoting sustainable economic growth.

Keywords: *neera*, beverage, preservation, coconut sugar, tapping

1. Introduction

Numerous traditional products and foods are derived from palms, including fresh juices and fermented beverages like toddy, wine, and arak, as well as concentrated syrup such as honey and brown sugar (jaggery powder). All of these are made by tapping the sap from various palm trees. Palm toddy is a fermented beverage popular among populations in Asia and Africa (Jirovetz et al., 2001). Palm sap, known as '*neera*,' is boiled for 3 to 4 h at temperatures above 100°C to produce brown sugar with a sweet flavour (Apriyantono et al., 2002; Ho et al., 2007). This sugar is commonly used in the production of chocolate, cakes, sweet soy sauce, food coatings, ice cream, and traditional Bengali sweets. Because of its sugar content, palm sap undergoes the Maillard reaction and caramelization during heating. The volatile compounds in coconut and palm sugar, including pyrazines, furans, ketones, fatty acids, and organic acids, contribute to their sweet, toasty, and nutty caramel-like aroma (Sohn and Ho 1995).

The sweet, translucent, and oyster-white sap extracted from the unopened inflorescence of coconut trees (Hebbar et al., 2022) is commonly referred to as '*neera*' in India, '*Raa*' in Sri Lanka, and '*Tuba*' in the Philippines. These drinks record traditional local names of '*toddy*' in Asia, '*mimbo*' in Cameroon, '*nsafufuo*' in Ghana, '*emu*'' in Nigeria (Jespersen 2003) or '*bandji*' in Côte d'Ivoire and Guinea (Karamoko et al., 2012).

Derived from the coconut blossoms before they mature into nuts, this sap is utilized as a natural bio-beverage and serves as a base for various nutritious food products. This sap is widely consumed as a natural nutritional beverage in regions such as Africa, Malaysia, Indonesia, Thailand, and Myanmar. Its popularity stems from its high nutritive value, agreeable flavour, and palatable taste. The sap is generally harvested in the early morning, and the tapping techniques employed differ between and within countries (Asha et al., 2019). Sugar is the primary component of fresh sap, and it has a low Glycemic Index (GI), and as a result, it has potential use as a therapeutic agent for managing type II diabetes mellitus (Devi et al., 2015).

Coconut inflorescence sap is highly nutritious and beneficial for the alcohol and vinegar industries (Ezeagu and Fafunso 2003). Among the various coconut products, it is particularly notable for its diverse by-products, which can generate substantial income (Cortázar et al., 2010). According to Barh and Mazumdar (2008), coconut sap is richer in nutrients than sugar palm juice and date palm sap. Kusumawaty et al., (2012) and Ysidor et al., (2015) highlighted that coconut sap has significant economic value, serving as a fresh drink, a fermented beverage, and as a raw material for sugar production in Southeast Asian countries such as Indonesia, Malaysia, India, and the Philippines. The sap contains 12–15% (w/w) sucrose, around 0.23% (w/w) protein, 0.02% (w/w) fat, and trace amounts of glucose, fructose, maltose, and raffinose, while being relatively rich in sodium and potassium ions (Kalaiyarasi et al., 2013). Furthermore, coconut sap contains significantly higher levels of vitamins and antioxidants compared to sugar palm and sugar cane juices (Asghar et al., 2020).

The natural sap serves as a source of indigenous microorganisms, including *Bacillus*, *Lactobacilli*, *Micrococci*, *Enterobacter*, *Leuconostoc*, *Saccharomyces*, *Candida*, and *Pichia* (Alonso et al., 2019). These microorganisms rapidly ferment *neera*, negatively impacting its palatability by producing astringency, unpleasant volatile compounds, and a cloudy appearance, leading to economic losses for *neera* producers (Arias and Murray 2012). To preserve and extend the shelf life of *neera*, heat preservation methods like pasteurization are employed. The National Chemical Laboratory in Pune, India has developed a specialized filtration technique to further enhance *neera*'s shelf life, while the Central Food Technological Research Institute (CFTRI) in Mysore, India and ICAR- Central Plantation Crops Research Institute (CPCRI), Kasaragod, India have developed technologies for its preservation and processing (Hebbar et al., 2022; Swamy 2013).

The exploration of *neera* as a value-added beverage and its transformation into various other products represents a promising avenue for enhancing the economic potential of coconut palm sap. This chapter will delve into neera processing, preservation, and product diversification, highlighting its nutritional benefits and market potential. By leveraging innovative methods and ensuring quality control, *neera* can be positioned as a healthy, natural alternative to conventional beverages, opening new opportunities for entrepreneurship and sustainable development within the coconut industry.

2. *Neera* as a Beverage

Neera is a widely recognized and valued product that is rapidly gaining popularity in the market. *Neera* is the sweet, non-alcoholic sap extracted from the inflorescence of the coconut palm, and is utilized primarily as a refreshing beverage to quench thirst (John and Ushadevi 2020). The coconut tree typically produces 12 to 14 inflorescences (spadix) annually, averaging one per month. Each spadix can sustain the development of 20-25 immature or mature coconuts. Immature coconuts require 6–8 months to reach maturity, while mature coconuts take approximately one year. At a water content of 500 ml per immature coconut, the total water yield from a spadix is estimated to be around 10-12.5 l. However, if the same spadix is tapped, it can produce approximately 60 to 67.5 l of sap over a period of 40 to 45 days, at a moderate yield of 1.5 l day^{-1} (Asha et al., 2019). Additionally, the sap is richer in nutrients and phytochemicals compared to coconut water. From a physiological perspective, tapping the spadix for sap is more energy-efficient than allowing it to develop into mature coconuts.

To prevent the development of sour or acidic tastes, the sap should be consumed within 12 h of collection. Following the fermentation of toddy, it can be distilled to produce arrack, a distinct alcoholic beverage with an alcohol content ranging from 33 to 50%. Other possible fermented products include coconut vinegar and vodka (Naik et al., 2015). Coconut inflorescence sap naturally ferments into toddy (Hebbar et al., 2018) with 4% alcohol within hours at ambient temperature, but can be processed and preserved in its unfermented form to retain its sugar, vitamins, and nutrients using advanced technologies. It is consumed as a natural nutritional beverage in countries such as Africa, Malaysia, Indonesia, Thailand, and Myanmar, gaining popularity due to its high nutritional value, pleasant flavor, and delicious taste. The unfermented sap is golden in colour, has a pH of 7.3 to 7.4, and is rich in nutrients (Singaravadivel et al., 2012). The time gap between sap exudation and concentration allows microbial activity, causing changes in sweet toddy sugars. The bark of *Vateria acuminata* is commonly used to reduce fermentation, but it imparts a harsh odor and undesirable color, making it unsuitable for beverage production (Gunawardhana et al., 2018). The ICAR-CPCRI, Kasaragod, India has developed and patented an improved method for collecting coconut sap in its unfermented state. This method, known as the Coco-Sap Chiller Method (CSCM), utilizes a simple ice-box system to collect the sap in a closed container maintained at a low temperature of 2-4°C (Hebbar et al., 2020).

The sap collected using the CSCM method remains unfermented and nutrient-rich, making it an ideal product for sale as a health drink. Additionally, value-added products such as sap concentrate and coconut sugar can be produced without the use of chemicals (Hebbar et al., 2015). Enhancing the value of unfermented coconut sap by creating flavored carbonated or non-carbonated ready-to-drink beverages presents a significant opportunity to promote it as a healthy and nutritious alternative to soft drinks in the market. Gunawardhana et al., (2018) investigated to develop an affordable non-alcoholic ready-to-drink beverage using unfermented coconut sap. They experimented with fortifying coconut sap by adding various fruit pulps, along with natural lemon flavor (0.56% v/v) and artificial mandarin flavor (0.8% v/v). The value-added flavored non-alcoholic ready-to-drink beverage can be successfully created from unfermented coconut sap by diluting it to 40% (v/v), adding 0.56% (v/v) natural lemon flavor, with or without carbonation, and applying thermal preservation at 95°C for 7 min.

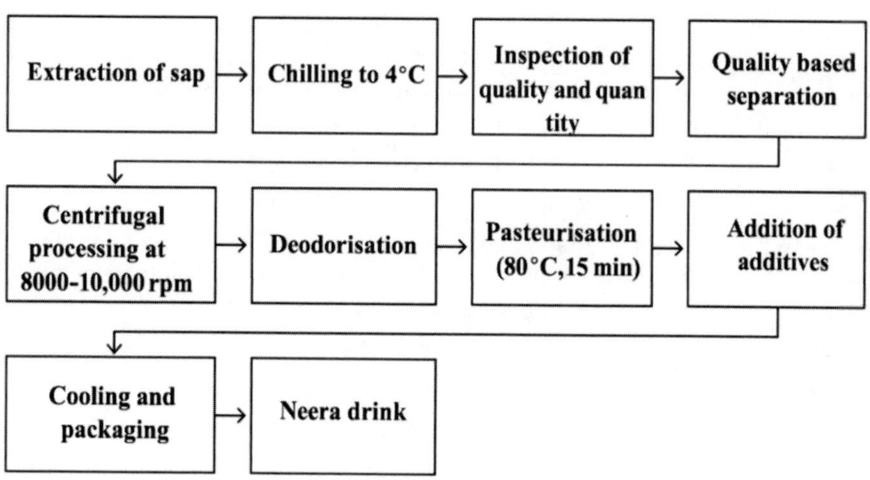

Figure 1. Preparation of sap as a drink.

Approximately 100 ml of sap delivers 75 calories of energy and 250 mg of protein, along with essential minerals, vitamins, and amino acids (Sarma et al., 2022). It is naturally abundant in magnesium, zinc, potassium, and iron. The sap contains about 16% carbohydrates, primarily composed of sucrose, with minor amounts of glucose, fructose, inositol, and raffinose. Fresh coconut sap is rich in two amino acids (glutamic acid and aspartic acid) naturally synthesized by the body, both of which offer numerous health benefits important for cognitive function and metabolism (Hebbar et al., 2020).

The Defence Research and Development Organization, India (DRDO) and the CFTRI, India collaborated to create a method for preserving coconut sap without its fermentation into alcohol (toddy). The process involves collecting sap from the unopened inflorescence of the spadix and immediately chilling it to 4°C. This is followed by filtration, further chilling to 2-8°C, adding citric acid (0.04-0.2%) as an acidulant, and nisin (10-15 ppm). Pasteurization at 95°C for 5-10 min is then applied for long-term storage. Additionally, centrifugation at 4000 rpm for 10 min is used to remove suspended particles (Asha et al., 2019) (Figure 1). Heat-processed *neera* was found to remain stable for 72 h. Although a one-year shelf life was reported when processed using in-pack pasteurization or retort pouch methods, and stored under refrigerated conditions, or for 30 days at ambient temperatures, it frequently developed undesirable taste and odour characteristics, making consumption challenging. Given that unfermented coconut sap is photosensitive due to its ascorbic acid content, packaging using polyethylene terephthalate (PET) combined with aluminum foil was developed for its storage. An alternative method was developed to preserve *neera* while maintaining its natural cloudiness and constituents. This process involves collecting the sap, filtering it, chilling it to 2–8°C, and adding bentonite or activated carbon black to remove undesirable odours. The sap is then stirred, centrifuged, filtered, and either pasteurized or treated with preservatives, resulting in neera that retains its natural cloudiness and has minimized odour (Asha et al., 2019).

3. Ultra High Temperature (UHT) Treatment

The coconut sap collected using the ice-box technology prescribed by ICAR-CPCRI and processed through Ultra-High Temperature (UHT) treatment retained most of its original qualities when stored under ambient conditions for the 103-day study period. Physical, biochemical, and mineral analyses of the UHT-treated *neera* indicated minimal impact of the process on its original quality, confirming that the treated *neera* is safe to store under ambient conditions for the duration of the study. Microbial investigations showed that the UHT-treated *neera* samples remained within the FSSAI-prescribed microbial limits for thermally processed juices, with no signs of spoilage throughout the shelf life study (47-103 days after treatment). The product remained microbiologically stable at ambient temperatures. Based on sensory evaluations and other analytical parameters, UHT-treated *neera* (125°C for 5 s) can be considered suitable for processing *neera* as a beverage (Ramesh et al., 2023). The product was launched commercially as *Thenneera Coco nectar* (https://thenneera.com/) (Figure 2).

Figure 2. A ready to serve (RTS) coconut inflorescence sap following ultra high temperature (UHT) treatment.

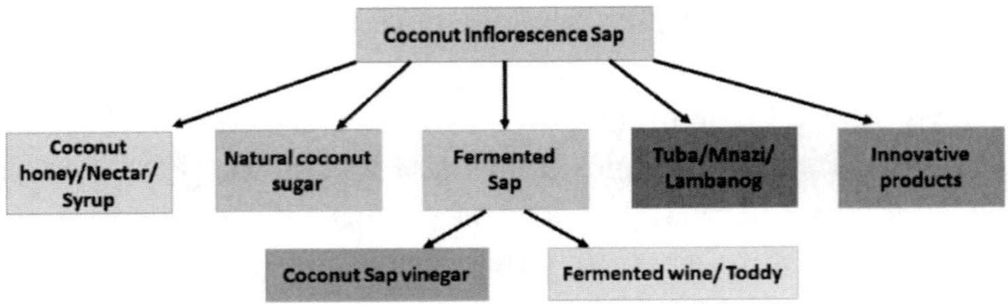

Figure 3. Value added products from coconut inflorescence sap.

4. Value Addition of Coconut Sap

Technological advancements in food processing are essential for addressing the increasing food demand, bridging the food security gap, and accommodating rising urbanization, which diminishes agricultural production while increasing food consumption. (Sharma et al., 2021). Coconut inflorescence sap-based products (Figure 3) like coconut sugar, jaggery, beverages, toddy, and vinegar are attracting considerable attention, primarily due to their health benefits. Most of these products are produced by small to medium-scale village-level operations. Coconut-based alcoholic beverages, including coconut arrack, coconut vodka, and coco wine, have also become popular in major coconut-producing nations.

5. Marketable Products of Coconut Sap

Coconut sap has the potential to be commercialized into a variety of products due to its nutritional and health benefits. It has been shown to possess several bioactivities, including antioxidant (Asghar et al., 2020; Devi et al., 2015), anti-inflammatory (Ratheesh et al., 2017), hepatoprotective, nephroprotective (Jose et al., 2017), antimicrobial, and cytotoxic (Rajesh et al., 2020) properties. Rich in nutritional components such as vitamins, minerals, amino acids, carbohydrates, proteins, and sugars (Table 1), the sap's content can vary significantly due to factors like coconut variety, harvesting (tapping, collecting, processing) methods, geographical location, and climatic conditions (Chinnamma et al., 2019; Sarma et al., 2021). Additionally, studies have found that the concentration of volatile compounds - such as acetic acid, dodecanoic acid, 1,4-dimethyl-6-1-butyl acetate, and 2-methylcyclohexane -

decreases from morning to evening during the tapping of coconut sap (Purnomo 2007).

Once collected, and if exposed, the coconut sap is susceptible to fermentation, which can diminish its commercial value. To inhibit fermentation and suppress microbial growth, the use of additives like limestone, potassium metabisulfite, and citric acid, as well as natural preservatives such as mangosteen skin, jackfruit bark, and chengal wood has been reported (Chinnamma et al., 2019; Purnomo 2007; Saidan et al., 2020).

The products that could be derived from coconut sap include Kalparasa (an unfermented drink), Rasgulla (a sugar-based dessert), coconut sap yogurt, and Coconut Sap Powder (CSP). CSP has shown an ergogenic effect and may be commercialized as an energy drink (Joseph et al., 2021). Furthermore, coconut sugar has a low GI value, making it an attractive option for diabetes and a better substitute for regular refined and muscovado sugars. It can also be incorporated into wheat-based products, such as bread made with coconut sugar, which has demonstrated acceptable quality compared to traditional bread made with sugar cane (Srikaeo and Thongta 2015). Coconut sap has also been utilized to enhance the shelf life and sensory characteristics of various foods, including fermented salted fish (Wattimena et al., 2021), acidophilus milk (Jirasatid and Nopharatana 2020), and animal feeds (Ravindran et al, 1984). Additionally, lactic acid bacteria (LAB) isolates derived from *neera* have shown probiotic qualities and antimicrobial activity, indicating their potential use as probiotics in food and feed formulations (Somashekaraiah et al., 2019).

6. Coconut Sugar

Coconut sugar, also referred to as coconut palm sugar, is a sweetener derived from the sap of the flower buds of the coconut tree. It is increasingly recognized as a natural alternative to refined sugar, offering various potential health benefits. Due to its low GI, coconut sugar has become a popular choice as a diabetes-friendly sweetener. It is commonly used in cooking, particularly for desserts and curries, and serves as a raw material in the food industry, including confectionery products. The production of coconut sap sugar relies heavily on key factors such as pH, temperature, and the level of microbial activity from the natural yeast present in the sap. Before boiling, the coconut sap should be strained to eliminate any unwanted materials. For successful granulated coconut sugar production, the pH of the sap should be maintained at above 6. However, drying coconut sap into coconut sugar powder can be challenging due to its high sucrose content (ranging from 12.30% to 17.40%), which can lead to stickiness during the drying process (Santoso 1993). There are two main types of coconut sugar: granulated coconut sugar, which is produced by continuously heating the supersaturated coconut sap until a crystalline structure forms; and molded coconut sugar, made by cooling the supersaturated sap in a special container to achieve a specific shape.

The sap initially contains 14-16% sucrose. When heated to 90-95°C, it forms long threads, which are then agitated or beaten to produce either crystalline or amorphous sugar. Coconut sugar is typically brown and has a moisture content of 2-3%. In contrast to commercially available sugars that are high in calories, palm-based sugars are rich in amino acids, vitamins, minerals, polyphenols, and antioxidants. One kilogram of sugar can be produced from approximately 7 l of sap. A coconut tree typically yields 400-500 l of sap annually, which equates to about 60-70 kg of sugar year^{-1}. A tree that produces 100 nuts each year provides around 38.5 kg of glucose from its economic yield, whereas tapping the tree to produce sugar generates about 133 kg of glucose, nearly 3.5 times more. Besides coconut sugar, a variety of secondary products can be made from this sugar as a base ingredient. Due to its health benefits, coconut sugar commands a premium price in both domestic and international markets. Consequently, tapping coconut trees and processing the sap into sugar not only benefits farmers but also creates significant employment opportunities for tappers, technicians, and sales personnel, while ensuring a high-quality product for consumers. Sap with a pH of ≥6.7 is ideal for making sugar (Hebbar et al., 2015). During heating, the sugars in the palm sap undergo Maillard reactions and caramelization. The volatile components of coconut and palm sugar - such as pyrazines, furans, ketones, fatty acids, and organic acids - contribute to their sweet, toasty, and nutty caramel-like aroma (Hebbar et al., 2018). The quality of *neera*, boiling temperature, and heating duration significantly influence the volatile components of palm sap (Martins et al., 2000). Both the yield and quality of the sugar are notably higher when high-quality raw material is used (with a pH around 7 and 14 to 16° Brix), whereas poor-quality *neera* results in a reduced sugar yield. Different processing methods are employed based on the type of heating and moisture extraction from the sap. Coconut sugar has a sweetness level comparable to sucrose, making it a suitable alternative sweetener for use in confectionery, pastries, beverages, and other culinary applications (Okoma et al., 2020).

7. Secondary Products from Coconut Sugar

7.1. Kalpa Bar

It is a chocolate made with coconut sugar and exclusively plant-based ingredients, free from dairy milk. Developed through a collaboration between ICAR-CPCRI and Mangalore (India) based Central Arecanut and Cocoa Marketing and Processing Cooperative Ltd. (CAMPCO), this chocolate includes cocoa powder, coconut sugar, natural vanilla extract, and GMO-free sunflower lecithin. It has low GI and contains no artificial additives. This dark chocolate is not only delicious but also maintains its form at room temperature, making it a great choice for those focused on health.

Table 1. Marketable products and nutritional values of coconut sap

Marketable products	Applications	References
Non-alcoholic products	Unfermented coconut sap drink/*Kalparasa/Neera*, energy drink (CSP)	Hebbar et al., (2015)
Fermented products	Toddy, vinegar, gluten-free soy source alternative	Mesquita et al., (2020)
Probiotic drinks	Lactic Acid Bacteria (LAB)	Somashekaraiah et al., (2019)
Preservative	Fermented-salted fish	Mahulette et al., (2016)
Milk emulsifier	*Acidophilus* milk, high-fat coconut milk	Jirapeangtong et al., (2008); Jirasatid and Nopharatana (2020)
Yoghurt	Coconut sap drink yoghurt	Karseno et al., (2021)
Sweetener	Syrup (Coconut Nectar, Sucrose heavy dessert e.g., Rasgulla)	Kaur and Goswami (2021)
	Coconut Sap Powder (COCOZENTM) Wheat-based foods	Srikaeo and Thongta (2015)
Chicken feed	Coconut sap distillery by-products	Ravindran et al., (1984)
Reducing agent	Synthesis of silver nanoparticles (AgNPs)	Joseph et al., (2021)
		Rajesh et al., (2020)
Fermentation medium	Levan production	Mummaleti et al., (2021)

Palm sugar is widely utilized in the production of chocolates and ginger toffee. To create these confections, a mixture of glucose and palm sugar is boiled at a temperature of 120°C. Once heated, the mixture is poured into trays and combined with powdered citric acid, along with a blend of color and essence. After thoroughly mixing the ingredients, the mixture is cut into slabs, allowed to cool, and then packaged as chocolate (Sharma and Meghwal 2021).

7.2. Kalpa Drinking Chocolate

This product combines coconut sugar with high-quality cocoa powder to create a rich and flavorful drinking chocolate. It dissolves easily in hot and cold milk, releasing a delightful chocolate aroma.

Source: https://bckv.edu.in/index.php/en/success-story-en.

Figure 4. Neera sugar or jaggery based milk sweets prepared in West Bengal, India.

7.3. Kalparasa-Based Milk Sweets

In West Bengal, India sweets are made using concentrated Kalparasa instead of other sweeteners (Ghosh et al., 2016). These sweets offer an alternative method for transporting *neera* over long distances. They are rich in minerals, vitamins, and valuable fibers not available in cane sugar-based sweets (Figure 4). *Neera* Sweet is a very fruitful confectionery product very much suitable for diabetic patients who love sweets. The sweet is made with kheer as a main ingredient and coconut sap as sweetener. The color of the sweet is yellowish and is named as Keshar Peda (Misra 2016).

7.4. Frozen Delicacy

ICAR-CPCRI has developed and commercialized a coconut-based frozen treat that uses only coconut ingredients, including coconut milk, coconut sugar, tender coconut water, and pulp. This premium product is both natural and healthy (Figure 5). The addition of coconut sugar enhances the flavor and taste of the delicacy (Beegum et al., 2022). It is suitable for individuals with lactose intolerance as it does not contain milk. Additionally, with refined sugar replaced by coconut sugar, the delicacy has low GI and high content of vitamins and minerals.

Source: ICAR CPCRI.

Figure 5. Coconut-based frozen delicacy.

7.5. Kalparasa-Based Nutribar

An energy-rich nutribar (Figure 6) can be made by concentrating Kalparasa to a thick consistency of 75–80 Brix (80% concentration). This is then mixed with protein-rich ingredients such as roasted peanuts and green gram, along with cocoa powder, and molded into bars.

Source: ICAR CPCRI.

Figure 6. Kalparasa-based nutribar.

7.6. Coconut Peda

A method for preparing peda using coconut milk and coconut sugar has been standardized (Figure 7). Use of 30% coconut sugar yielded the best results, with a recovery rate of 50–53%. The peda made with coconut sugar was more acceptable than the peda made with refined sugar. Besides, the peda made with coconut sugar had higher mineral (ash) and protein contents (Manikantan et al., 2020).

7.7. Instant Kheer Mix

An instant kheer mix based on coconut can be made using coconut milk powder, vermicelli, and coconut sugar in a ratio of 1.5:1.5:1 (Figure 8). The standardized water-to-mix ratio for reconstitution is 6:1. Analysis of quality parameters such as moisture content, flowability, bulk density, particle density, interstitial air content, pH, titratable acidity, and free fatty acids over three months period confirmed the product's shelf life.

Source: ICAR CPCRI.

Figure 7. Coconut peda.

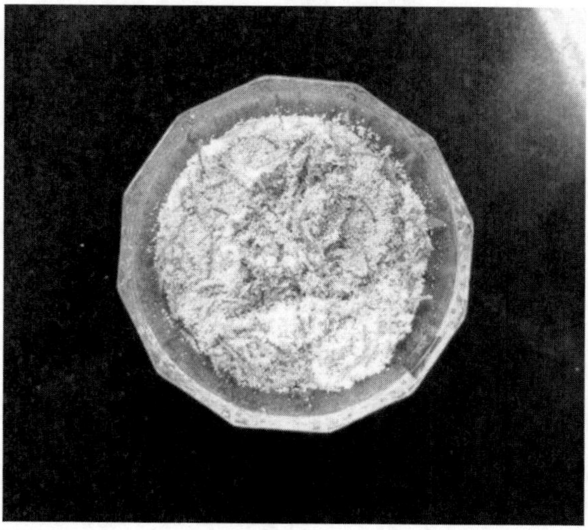

Source: ICAR CPCRI.

Figure 8. Instant kheer mix.

7.8. Coconut Bites

The Central Food Technological Research Institute (CFTRI) in Mysore, India has developed technology for value-added coconut products, specifically Coconut bites, a ready-to-eat sweet snack. Coconut Bites retain the distinct flavor of fresh coconut and are available in various colors and flavors to cater to consumer preferences. The production process involves washing and dicing the coconut, cooking the pieces in sugar syrup for 45 min. and then removing them from the syrup for drying in a hot air oven. Afterward, the dried cubes are coated with sugar syrup in a coating pan, where approved food colors and flavors are applied. The initial moisture content of Coconut bites was 2.6% (at 27°C, 65% RH). The product was then packaged in 200-gauge pouches (50-gauge metalized polyester/150-gauge polyethylene) and stored at room temperature for up to 4 months (Muralidharan and Jayashree 2011).

7.9. Jaggery-Infused Osmo-Dehydrated Coconut Chips

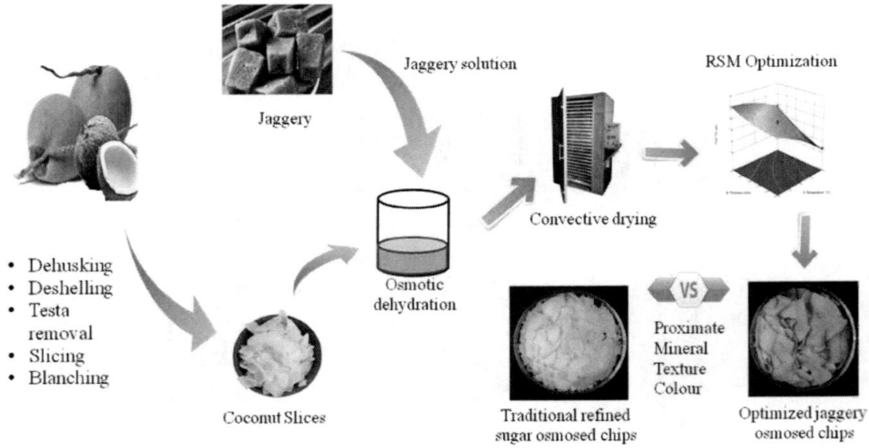

Figure 9. Jaggery infused osmo-dehydrated coconut chips (Pravitha et al., 2021).

Coconut slices can be transformed into a healthier snack by using jaggery as an alternative to conventional osmotic agents, thereby avoiding the negative health impacts associated with refined sugar. The comparison between the newly developed product and traditional coconut chips showed that jaggery serves as an excellent alternative to non-nutritive osmotic agents, resulting in a mineral-rich snack with similar textural properties (Figure 9). These findings could be valuable for researchers and entrepreneurs focused on adding value to food products and developing new ones. Additionally, the use of jaggery as an osmotic agent in food product development has not been extensively explored, indicating a need for further research to investigate its potential in various fruit processing applications to enhance the health benefits of the final products (Pravitha et al., 2021).

8. Fermented Coconut Inflorescence Sap for Virgin Coconut Oil Production

The research study conducted by Pandiselvam et al., (2023) focused on optimizing the production process for virgin coconut oil (VCO) by investigating the use of aerobically fermented coconut sap. Fermented *neera* can be made directly from fresh *neera* without any extra costs, offering a potential advantage compared to commercially available cultures. Coconut milk was extracted from the coconut kernel using a hydraulic expeller. This milk was then fermented with different concentrations of fermented *neera*, ranging from 5 to 25%, and the oil layer was subsequently separated. Use of fermented *neera* as a starter at 10% concentration produced the highest VCO yield of 19.8% after 14 h of fermentation. The fermentation time of the starter (6 h, 8 h, and 10 h fermented *neera*) and its concentration significantly influenced the VCO yield. The physicochemical characteristics of the VCO produced through the fermented *neera*-based coconut milk fermentation method complied with the standards set by the Food Safety and Standards Authority of India (FSSAI) and the Asian and Pacific Coconut Community (APCC). Therefore, using fermented *neera* presents a new and cost-effective approach to accelerate the fermentation process in VCO production.

8.1. Tuba

The coconut inflorescence sap collected in Colima, Mexico, is referred to as 'tuba,' also known as 'nectar' or 'honey water' (Chandrasekhar et al., 2012; Kadere and Kutima 2012; Monreal et al., 2011; Sánchez and Ríos 2002). In the Philippines, however, 'tuba' (Figure 10) specifically denotes the alcoholic beverage produced by fermenting fresh sap, which is also called toddy in Sri Lanka, tuak in Indonesia, legmi in Tunisia, mnazi in Kenya, emu in Nigeria, nsafufuo in Ghana, mimbo in Cameroon, and kallu in India (Chandrasekhar et al., 2012; Gallegos et al., 2019).

(a) (b)

Figure 10. Samples of tuba: The one on the left (a) is only in its first few days but the one in the right (b) has been aged for months or years (Agaton 2022).

This beverage is popular in many Southeast Asian countries, and is commonly consumed in the coastal regions of Colima, Michoacán, and Guerrero, Mexico, where it is often enjoyed mixed with nuts (Gallegos et al., 2019). In all these instances, the drink is created through spontaneous fermentation, which encourages the growth of yeasts and bacteria that convert the sugar-rich coconut sap into an alcoholic beverage containing various nutritional components, such as amino acids, proteins, vitamins, and sugars.

Its production involves spontaneous fermentation by various microorganisms, including LAB that produce bacteriocins, along with mesophilic aerobic bacteria and yeast (Singhal et al., 2008). The tuba collected in the morning (between 6 and 7 am.) is known as sweet tuba. It has a thick, transparent, and highly viscous appearance, making it extremely sweet. In contrast, the tuba collected in the afternoon, which drips throughout the day and is subjected to higher temperatures, is less viscous and not as sweet. Generally, fresh sap is sweet, oyster-white, and translucent, with a neutral pH. However, its composition and quality can vary based on factors such as location, time of collection, and the duration of tapping. Locals empirically use tuba as a remedy for gastrointestinal issues and as a rehydration drink (Monreal et al., 2011; Salcido et al., 2015). The production of tuba can be optimized by controlling the presence and abundance of beneficial microorganisms that contribute to its functional characteristics (Melgar et al., 2019).

According to local producers, tuba is well accepted in Colima, Mexico, where various fruits and seeds, such as strawberries, apples, cucumber, and peanut seeds, are added for flavor. The color of tuba can change due to the action of microorganisms, with fermentation beginning as soon as five hours after harvesting. When left to rest for about eight days, tuba can be transformed into a useful vinegar for domestic purposes. It is reported that tuba contains sugar levels of up to 90 g/l, which are fermented by microorganisms, along with vitamin C, phosphorus, amino acids, and minerals (Monreal et al., 2011).

Tuba is primarily a fermented drink made from coconut sap mixed with barok, the reddish bark of the mangrove tree, which is dried and chopped into pieces, giving the drink its red color (Agaton 2022). To prepare tuba, the collected coconut sap is transferred to a *kawit*, a large container typically made of bamboo, and is mixed with barok. The quality and quantity of barok used are critical, as they significantly affect the final quality of the tuba. Fresher barok yields a better flavor, while a higher quantity of barok results in a deeper red hue. However, adding too much barok can create a very bitter taste, while too little leads to an imperfect mixture. After mixing, the tuba is placed in a larger container, often a plastic jug. Glass containers are preferred for aging tuba, as plastic containers can sometimes impart an undesirable taste, especially if they are new. This plastic-like flavor is typically absent in glass containers. In Palo, Leyte, Philippines, some people bury tuba underground to facilitate the aging process, while others add raisins to enhance its aroma.

During the aging process, tuba should not be exposed to direct sunlight, as this can cause it to turn into vinegar. Additionally, the containers must be filled to the brim; otherwise, the aging process may be compromised, potentially leading to vinegar formation. In the early stages of fermentation, sediment forms at the bottom of the container due to the distillation process. To ensure quality, repeated distillation is necessary until no more sediment accumulates.

Tuba is referred to as bahal when aged for six months and as bahalina when aged for 12 months or longer. The alcohol content increases with extended aging (Sanchez 2009). The tuba remains in the container until it reaches the bahalina stage, but it is ready for consumption at any point during the fermentation process. Some people prefer to drink it early when it is still sweet, while others wait until it has matured into bahal or bahalina. Coconut sap without barok typically has an alcohol content of 2-4%, while sap combined with barok and aged further can contain 10-14% alcohol.

The presence of bacteriocin-producing bacteria in fermented beverages is crucial as they protect these foods from microbial contamination and spoilage, enhance flavor and aroma, and provide indirect protection against potential pathogenic bacteria for consumers. Salcido et al., (2015) demonstrated the presence of potential probiotic and bacteriocin-producing LAB such as *Lactococcus lactis* TuAB1, *Lactococcus lactis* TeA1, *Enterococcus faecium* TuAB2, and *Enterococcus faecium* TeA2 in Mexican tuba. These bacteria are part of the cultivable microbiota and can exert inhibitory effects against clinically significant pathogenic bacteria.

8.2. Mnazi

Mnazi is a whitish, effervescent, and acidic tropical alcoholic beverage produced by fermenting sugary coconut sap. This drink results from a combination of alcoholic, lactic, and acetic fermentation. If not preserved, mnazi can turn into vinegar (acetic acid) within two to five days due to continuous fermentation, making vinegar one of its valuable by-products. Initially, the sugars in the sap are fermented into ethanol within 8 to 12 h by yeasts and LAB, creating a conducive environment for the growth of acetic acid bacteria. These bacteria typically appear two to three days into the fermentation process, utilizing glucose and/or sucrose, which may also be present at earlier stages of mnazi fermentation (Okafar 1975).

Mnazi goes by various names and has numerous regional variations, including emu and ogogaro in Nigeria, tuba in the Philippines, ra or panam culbo in Sri Lanka, tuak or nira in Malaysia, nsafufuro in Ghana, kallu in India, and toddy in Thailand. In Kenya, it is commonly referred to as mnazi, uchi wa mnazi, or mdafu. In West Africa, mnazi is often called 'Nigerian wine' (Faparusi 1971).

Coconut wine is free from methanol and meets the standards set by the Kenya Bureau of Standards for fortified wines (KS112:2007), table wines (KS 05-609:1990), and sparkling wines (KS 05-1121:1994) (Kadere et al., 2004). It is safe to drink and has the potential to capture 40% of the domestic alcoholic beverage market in Kenya with adequate support and development. According to extensive surveys conducted in Kenya, Kadere et al., (2009) reported that earnings from mnazi surpass those from other coconut-based products in many districts.

Kadere et al., (2008) identified *Acetobacter* and *Gluconobacter* as the dominant spoilage genera of acetic acid bacteria in coconut wine (mnazi). These bacteria appear two to three days after the fermentation of coconut sap begins, and their population can affect the composition of mnazi, the growth of yeasts during alcoholic fermentation, and the proliferation of LAB during malolactic fermentation. Kadere and Kutima (2012) noted the presence of LAB in mnazi as spoilage organisms and suggested that controlling these bacteria is essential for the effective utilization of mnazi.

In a study examining the quality and preservation of mnazi and its spirit (pyuwa), Kadere (2021) found the main volatile compounds present to be propanol, isoamyl ethanol, butanol, and acetic acid. In mnazi, the levels of these volatiles were significantly below 600 mg l^{-1}, which is considered the threshold of acceptability for wine. The absence of methanol and fusel oils (volatile, oily liquids produced in small amounts during fermentation) in fresh mnazi allows it to compete effectively with existing beers and alcoholic beverages in the market. However, its distillate (pyuwa) is not recommended as a safe alcoholic drink unless further separation is performed due to its high levels of fusel oils. Value-added products such as dry, medium dry, and sweet brands prepared from mnazi remained stable during the first four weeks after production. The assimilation of sucrose and fructose occurs more rapidly than that of glucose, with fructose being the quickest to ferment.

8.3. Lambanog (The Philippine Vodka)

Coconut wine, commonly referred to as Lambanog and affectionately known as the "Philippine Vodka," is a natural and chemical-free alcoholic beverage produced through the fermentation and distillation of naturally sourced coconut sap (Velasco 2013). It is renowned for its high alcohol content, typically a minimum of 30%, and its potency. Lambanog serves as a base liquor for creating flavored spirits and cocktails, and it holds a special place in Filipino culture, being integral to festive celebrations, social gatherings, and special occasions.

Lambanog is widely produced across various regions in the Philippines, with the Southern Tagalog Region being the primary production area. The brand Lakan, a Philippine lambanog, has garnered international recognition, winning medals in all eight international wine and spirit competitions, including two gold medals at Europe's prestigious Monde Selection in 2015 and 2017, as well as a double gold in a Chinese competition. These international accolades underscore the reputation of genuine lambanog as a noteworthy beverage.

Coconut lambanog is classified into two categories: pure and flavored. Pure lambanog is unmodified, clear, and colorless, and is commonly found in local stores. In contrast, flavored lambanog includes the addition of fresh fruits or FDA-approved food colors and flavors, appealing to more adventurous drinkers.

The Philippine National Standard for Distilled Fermented Coconut Sap (Coconut Lambanog) specifies that coconut lambanog is a colorless distilled spirit obtained by distilling naturally fermented coconut sap, which is collected and allowed to ferment for 48 to 72 h before distillation. This sap is the liquid that oozes from the tapped unopened inflorescence of the coconut palm, also referred to as "coconut nectar" in the Philippines.

Flavored coconut lambanog includes natural and/or artificial flavors and/or permitted food colors as approved by the Food and Drug Administration (FDA). For details on essential composition and quality factors of coconut lambanog, please refer to Table 2. The various *neera* based drinks in the Philippines and Thailand are presented in Figures 11 and 12.

Table 2. Essential composition and quality factors of coconut lambanog

Physico-chemical characteristics	Levels
Alcohol content, expressed as ethyl alcohol, minimum	30%
Methanol	The methanol content shall be in accordance with the provisions of BFAD M.C. No.13 s.1989: Methanol may be present in alcoholic beverages provided that it shall be derived from the natural alcohol fermentation process and not added.
Titratable acidity, expressed as acetic acid, maximum	0.3%

9. Coconut Vinegar

A natural alternative to synthetic vinegar, coconut vinegar can be produced from either coconut water or the sap of the coconut tree. It is widely used as a flavoring agent and preservative and is rich in vitamins and minerals. The acidification process continues for up to 7 days, after which the coconut vinegar is harvested by decanting (Mithra et al., 2013).

Coconut vinegar is produced by pouring fresh coconut sap into a large plastic container covered with a clean netted lid, allowing for aeration while preventing contamination from dirt and other foreign materials. After a fermentation period of 10 days in a well-ventilated environment, the coconut sap can be harvested as vinegar. To ensure quality, the vinegar is pasteurized by heating it to 60–65°C for 5–10 minutes, which maintains a minimum acidity of 4%. Once it has cooled sufficiently, the vinegar is poured into very clean bottles and sealed tightly.

Figure 11. Coconut sap drink in the Philippines (Photo credit: Deepti Nair S).

Figure 12. *Neera* in Thailand (Photo credit: Deepti Nair S).

Coconut vinegar serves various purposes, including its extensive use as a preservative in the pickle industry and as a flavoring agent in food processing. Compared to synthetic vinegar, palm vinegar has strong export potential. The fermentation process involves the yeast and acetobacter naturally present in the sap, where yeast first ferments the sugars into alcohol, which is subsequently converted into vinegar by acetobacter (Chandrasekhar et al., 2012).

10. Palm Wine

Palm wine is an alcoholic beverage produced from the sap of young inflorescences of both male and female palms of various species such as the palmyra, date palms and coconut palms. The sap starts to ferment immediately after its collection, due to natural yeasts in the pores of container pot and air (often spurred by residual yeast left in the collecting container).

If allowed to ferment for more than a day, its alcohol content can reach up to 4%, which is mildly intoxicating, sweet and it eventually starts converting into vinegar (Chandrasekhar et al., 2012). Fermentation of wine for a day yields a stronger, more sour and acidic taste. Oftentimes, it takes a few weeks before the sap can even be called wine. Afterwards, it becomes something completely different based on its processes and length of fermentation.

The terms 'toddy' and 'palm wine' both refer to similar alcoholic beverages, though the terminology can vary by country. In some parts of India, the unfermented sap is known as '*neera*.' In India, palm wine or toddy is served as either *neera* or *padaneer* (a sweet, non-alcoholic beverage derived from fresh sap) or *kallu* (a sour beverage made from fermented sap, but not as strong as wine). *Kallu* is usually drunk soon after fermentation by the end of day, as it becomes more sour and acidic day by day. Sometimes, spices are also added to give a distinct taste.

In India, mainly three types of palm wines are available i.e., palmyra palm wine, silver date palm wine and coconut palm wine. The sap extracted from incising coconut flower clusters is consumed as *neera* and fermented to produce coconut palm wine, known as '*tub*' in the Philippines. Spices are also added during the brewing process to enhance the drink's distinct flavor. This sap can also be boiled down to create a sweet syrup or candy. One liter of palm wine contains around 300 calories, 0.5-2.0 g of protein, and significant amounts of vitamins, particularly vitamins A, C, and K, which help protect and improve eyesight (Swamy 2013). Palm wine acts as a starter culture for various fermentations. In Asia, it is used as an inoculum for cottage industry fermentations like Nan, a traditional leavened bread, and Sonnon, a preparation of steamed rice flours mixed with palm wine.

Njugu (2013) examined the competitiveness of the coconut wine value chain within the coconut industry in Kenya, noting that key products in the toddy value chain include fresh wine or coconut juice, fermented wine, coconut sugar, and vinegar. Coconut wine can be utilized to create various products, such as alcoholic coconut wine (mnazi), coconut syrup (a non-alcoholic beverage), coconut sugar, vinegar, acetic acid, coconut sap drinks, and spirits, among others. The coconut wine produced in Kenya complies with the Kenya Bureau of Standards for fortified wines (KS112:2007), table wines (KS 05-609:1990), and sparkling wines (KS 05-1121:1994). With no traces of methanol, coconut wine is safe for consumption and has the potential to capture 40% of Kenya's domestic alcoholic beverage market if adequately supported and developed.

In a related study, Udomsaksakul et al., (2018b) isolated and identified 26 strains of *Saccharomyces cerevisiae* from coconut inflorescence sap in Thailand. Among these, strain NL010 produced coconut wine with an alcohol content of 10.3% (v/v), exhibiting a satisfactory overall acceptance and flavor profile. This indicates that coconut inflorescence sap could serve as a viable source of indigenous *S. cerevisiae* strains for coconut wine fermentation.

11. Coconut Sap Syrup

Coconut sap syrup is created by heating fresh sap until it thickens into a syrupy consistency. The resulting syrup has a golden hue, and the recovery rate from the sap used is approximately 16%. This syrup is rich in iron, making it beneficial for individuals with anemia. In various countries, palm syrup is recognized as a health and wellness drink and is often included in Ayurvedic and other pharmaceutical formulations.

12. Innovative Products from Coconut Sap

In Banyumas Regency, Central Java, Indonesia, the utilization of coconut sap by medium and small micro enterprises (MSMEs) is primarily limited to coconut sugar and crystal sugar production. Ahsan et al., (2019) proposed the potential development of 30 different coconut sap products, categorizing them based on suitability for MSMEs in Banyumas Regency:

- Suitable Category: White crystalline sugar, fresh drinks, and jelly drinks.
- Quite Suitable Category: Syrup, candy, soy sauce, nectar, yakult, and yogurt.
- Not Suitable Category: Kefir.

The juvenile inflorescence of the coconut palm produces non-alcoholic palm nectar, which serves as a bio-refreshing drink and a raw material for various value-added products, including honey, syrup, jaggery, and sugar through boiling and thickening of the fresh sap (Rajath and Santhoshkumar 2020).

Sudana and Waisnawa (2020) studied the impact of temperature (85, 90, and 95°C) and heating duration (0, 10, 20, and 30 min) on the alcohol content of coconut sap. They found that the average alcohol content decreased as heating duration increased, with ranges of alcohol content as follows:

- At 85°C: 63% to 72%
- At 90°C: 49% to 56%
- At 95°C: 22% to 34%.

12.1. Spray Dried Coconut Sap Powder

Spray drying is employed to produce unfermented coconut sap powder, extending its shelf life. Yoghapriya and Sivakami (2011) achieved high-quality coconut sap powder through this method, resulting in a pale brownish powder with a slightly sticky texture due to its high sugar content (14 mg). The powder had a sweeter taste and better flavor attributes compared to palm *neera* powder. Jayasundera and Kulatunga (2014) reported a maximum recovery of 56.35% through spray drying of unfermented coconut sap (sweet toddy) mixed with maltodextrin in an 80:20 ratio. The resulting powder had a moisture content of 1.24% and a water activity value of 0.29%, maintaining quality for 12 months. The powder was amorphous with spherical particles, containing a total sugar content of 66.1%, which included 32.7% sucrose, 22.4% glucose, and 11.0% fructose.

Table 3. Physico-chemical characteristics of coconut inflorescence sap powder (Joseph et al., 2021)

Parameters	Results
Colour and appearance	White free flowing granular powder
Taste	Sweet
Bulk density (g ml^{-1})	0.2–0.5
Moisture (%)*	4.8
Water activity*	0.25
Ash content (%)*	0.95
Solubility	Soluble in water
Protein(%)*	1.5
Fat (%)*	1
Carbohydrate (%)*	78

*Official Methods of Analysis of AOAC International, 19th Ed.

Yuvarani et al., (2020) adjusted the total soluble solids (TSS) of unfermented coconut inflorescence sap with maltodextrin as a carrier agent at different concentrations (20, 25, and 30° Brix) and inlet air temperature levels (130, 140, and 150°C). They found that an inlet air temperature of 130°C with 25° Brix maltodextrin produced optimal drying characteristics, achieving a recovery of 198 g L^{-1} of sap used, with a moisture content of 2.81% and excellent flowability. Maltodextrin aids in minimizing powder deposition in the drying chamber and balances hygroscopic and hydrophobic properties to alleviate stickiness in coconut sap powder preparation (Nurhadi et al., 2018). The coconut inflorescence sap powder (COCOZEN™) has a GI of 52.47 and is recognized for its safety and acceptable sensory attributes (Mammen et al., 2020). The physico-chemical characteristics of coconut inflorescence sap powder are summarized in Table 3.

12.2. Ready-to-Drink Beverage

Chilled coconut sap is a highly nutritious beverage that offers a variety of essential nutrients, including minerals, vitamins, and proteins. Notably low in caloric content and GI, this drink is both sweet and delicious, making it an appealing choice for consumers. Studies suggest that coconut sap can be a superior alternative to mineral water, particularly as a thirst-quencher. Its natural, non-alcoholic nature and rich nutrient profile have led to a growing demand for this refreshing drink.

To ensure the freshness of the sap and prevent natural fermentation, the freshly harvested coconut sap undergoes pasteurization for 10 min at a temperature of 60-65°C. After pasteurization, the sap is kept for cooling before being transferred to suitable containers, which should be sealed tightly. For optimal preservation, it is recommended to store the sap in the cool section of a refrigerator or freezer. With proper hygiene practices during preparation and handling, the sap juice can be stored for up to three days without deterioration.

Gunawardhana et al., (2018) successfully developed a carbonated, flavored, non-alcoholic ready-to-drink beverage using unfermented coconut sap. The formulation included a 40% (v/v) dilution of the sap, with 0.56% (v/v) natural lemon flavor added for taste enhancement. The beverage underwent thermal preservation at 95°C for 7 min ensuring microbiological safety. The resulting beverage demonstrated excellent shelf stability, remaining microbiologically safe for up to 120 days when stored at refrigerator temperature (4 ± 2°C). This innovation highlights the potential of coconut sap not only as a traditional drink but also as a versatile ingredient in modern beverage production, catering to the growing market for health-conscious consumers.

12.3. Coconut Sap Powder

Coconut sap powder (CSP) has emerged as a novel product through a recently developed low-temperature preparation method that effectively preserves both macro- and micronutrient quality (Joseph et al., 2021; Mammen et al., 2020).

This innovative processing technique is significant as it not only enhances the nutritional profile of coconut sap but also contributes to its functional health benefits. Research has shown that CSP can mitigate oxidation-induced changes in human peripheral blood mononuclear cells. Furthermore, it has been linked to an increase in circulating antioxidant biomarkers in young healthy adults and a reduction in oxidative and inflammatory markers associated with kidney disease in preclinical animal models (Joseph et al., 2021; Ratheesh et al., 2017; Renjith et al., 2013). This suggests that CSP may play a role in combating oxidative stress and inflammation, which are critical factors in various chronic diseases. The composition analysis of CSP reveals that it is rich in several essential vitamins, particularly B vitamins and vitamin C, as well as important minerals and micronutrients such as manganese, calcium, iron, and potassium. Additionally, it contains flavonoids and nitrates, compounds known for their potential to improve arterial function (Filippini et al., 2020; Haddy et al., 2006; Rossman et al., 2021). This nutrient-rich profile makes CSP a valuable addition to the diet, particularly for individuals seeking to enhance their overall health. In a recent study, the consumption of coconut sap powder at a dosage of 1.5 g day^{-1} for eight weeks resulted in lowered brachial systolic blood pressure and reduced mechanical stiffness of the carotid artery in middle-aged and older adults (Carlini et al., 2023). These findings indicate a potential cardioprotective effect of CSP, highlighting its role in supporting cardiovascular health. The development and consumption of coconut sap powder offer numerous health benefits, including antioxidant properties, anti-inflammatory effects, and potential cardioprotective effects. Its rich nutrient profile further enhances its appeal as a functional food product. As research continues to explore the benefits of CSP, it may become a staple in health-conscious diets, contributing to improved overall wellness.

12.4. *Neera* Concentrate

Chinnamma et al., (2019) developed a protocol for preparing coconut sap concentrate (*neera* concentrate). As a natural drink with 80% water and all primary and essential nutritional constituents, 1000 ml of coconut sap can be concentrated to 200 ml as a final product. Based on its texture, it can be diluted to prepare a natural *neera* drink by dissolving in water without adding sugar. Such *neera* concentrate can be used as a bread spread and for making cakes, jilebi, ladoo, biscuits, bread, etc.

12.5. Cellulose

Ultrafiltration membrane is widely used in separation processes, and cellulose acetate is one of materials used for the production of ultrafiltration membrane. Coconut sap is a conducive medium for microorganisms to grow so that the process of glucose conversion into cellulose by *Acetobacter xylinum* bacteria is higher. Vaulina et al., (2018) were successful in the production of ultrafiltration membrane using coconut inflorescence as a source of bacterial cellulose. Widyaningsih et al., (2020) through a fermentation process, used coconut sap as a basic material in the production of cellulose acetate membranes. Such cellulose acetate membrane can be applied in the process of desalination.

12.6. Coconut Sap Amino

Coconut aminos is a dark brown, salty sauce made by fermenting coconut sap, similar to soy sauce in taste and consistency. After extraction, the sap is stored and aged. During this time, it ferments due to the natural sugars present within it. Some salt, coconut nectar and sugar are also added. Coconut aminos is a popular alternative to soy sauce since it is gluten-free, soy-free, and low in sodium, but it still adds a lot of flavor. It also has high amino acids, vitamins, and mineral contents. It has antioxidant, anti-inflammatory, and antimicrobial properties and provide a healthier alternative to soy sauce. Coconut aminos tastes similar to soy but is slightly sweeter. Despite its name, it does not taste like coconut but adds umami, or a savory flavor, to foods. Coconut aminos contains less salt than soy sauce, and the Na content varies widely between brands, from 66-160 mg $5ml^{-1}$ teaspoon, whereas soy sauce contains about 300 mg $teaspoon^{-1}$.

12.7 Carbon Dots

Carbon dots (CDs) are a new type of carbon-based nanomaterial that have attracted intensive interest from researchers, all over the world, due to their excellent biophysical properties. Murugesan et al., (2020), for the first time, selected coconut sap as the source for the synthesis of fluorescent CDs using the hydrothermal method. They found *neera* derived CDs as a promising application for the detection of Ag+ ions in water samples. Thus, the synthesized CDs can be used as an eco-friendly material in fluorescent probes for detection of metal ions in aqueous solution.

12.8. Yoghurt

Coconut sap drink yogurt is a fermented product based on coconut sap and skim milk made by fermentation using LAB of *Lactobacillus bulgaricus* and *Streptococcus thermophilus*. Karseno et al., (2021) prepared such yogurt using a combination of 10% skim milk and 50% of coconut sap having similar characteristics of commercial yogurt.

12.9. Levan

The levan is a natural microbial exopolysaccharide that gained intensive interest in commercial production and various industrial applications. It finds usage in gum, emulsifier, stabilizer, thickener, encapsulating agent, and a perfect raw material for the production of green plastics (Wu et al., 2013). It can be used as an alternative for synthetic sweetener. Mummaleti et al., (2022) studied the potential usage of fresh coconut inflorescence sap as a fermentation medium for production of levan using *Bacillus subtilis*. The maximum levan produced was 51.84g l^{-1} at the optimised conditions of temperature – (35°C), pH (6.5), and agitation speed (150 rpm). In the fed-batch fermentation process, there was an increase in levan yield to 62.1g l^{-1}. The maximum levan was obtained at pH of 10 with ethanol at 1:5 ratio.

12.10. Ethanol

Among the ten ethanol fermenting yeast species such as *Saccharomyces cerevisiae, Lachancea fermentati, Pichia kudriavzevii, Shizosaccharomyces pombe, Candida tropicalis, Zygosaccharomyces rouxii, Saccharomycodes ludwigii, Hanseniaspora guilliermondii, Wickerhamomyces anomalus and P. manshurica* isolated from coconut inflorescence sap in Thailand (Udomsaksakul et al., 2018a), *S. cerevisiae* was the dominant species (65%). The strain NT029 gave the highest ethanol yield (64.5 g l^{-1}), while the strain NC018 had the desired characteristics of a baker's yeast strain (high CO_2

production, high biomass yield and fast flocculation), and the bread prepared using this strain showed a good texture.

13. Challenges in Advancing *Neera*

Neera, which is non-alcoholic, is currently categorized as toddy under the Abkari Act, Govt. of Kerala, India even though toddy is a fermented drink with an alcohol content of 7 to 9% and is considered liquor. This classification is inappropriate, as *neera* does not come under the purview of the definition of toddy or liquor. Unlike toddy, *neera* is not subject to the Excise Act in coconut-producing countries, and it should be treated as a distinct product. The marketing of processed and packaged *neera* also should not fall under the prevailing Abkari Act.

One major issue is the lack of available palms for tapping, as coconut farmers receive insufficient lease payments, making it economically unviable for them. Additionally, the shortage of skilled tappers trained in the scientific extraction and processing of *neera* presents another challenge. The production volume is heavily reliant on the expertise of the tappers.

The Coconut Development Board (CDB), Govt. of India has proposed the formation of a specialized workforce known as '*neera* technicians' to handle *neera* tapping, which could generate new 'green collar job' opportunities. The marketing of *neera* and its value-added products can be pursued through innovative strategies (Kochukalam 2014).

Conclusion

Coconut *neera* is a versatile beverage with significant potential due to its natural sweetness, rich nutrient content, and health benefits. Beyond its use as a refreshing drink, coconut *neera* can be transformed into a variety of value-added products, such as jaggery, syrup, vinegar, and fermented beverages. Its adaptability makes it a promising ingredient in the food and beverage industry, offering both economic and health advantages. However, the general public remains filled with speculation and doubt regarding *neera*. When fermented, *neera* transforms into toddy. Coconut farmers, who have long allowed their palms to be tapped, have always recognized the health benefits of *neera* and have traditionally consumed it in its unfermented form. As consumer demand for natural and functional foods continues to grow, coconut *neera* and its derivatives present significant opportunities for innovation and commercialization in the food industry. Continued research and development will be essential to optimize production processes and expand the range of applications for this valuable resource.

References

Agaton, S. I. G. (2022). The tuba culture in Leyte, Philippines. *Recoletos Multidisciplinary Res J*, *10*(2). https://doi.org/10.32871/rmrj2210.02.01.

Ahsan, A., Mela, E., Widjonarko, G., & Sitoresmi, I. (2019). Produk nira kelapa untuk pengembangan usaha mikro, kecil, dan menengah di Kabupaten Banyumas (Products of coconut nira to develop medium and small micro enterprise in Banyumas Regency). *Agrin*, *23*(2), 85–102.

Alonso, S., Castro, M. C., Berdasco, M., de la Banda, I. G., Moreno-Ventas, X., & de Rojas, A. H. (2019). Isolation and partial characterization of lactic acid bacteria from the gut microbiota of marine fishes for potential application as probiotics in aquaculture. *Probiotics and Antimicrobial Proteins*, *11*, 569–579.

Apriyantono, A., Aristyani, A., Lidya, Y., Budiyanto, S., & Soekarto, S. T. (2002). Rate of browning reaction during preparation of coconut and palm sugar. In *International Congress Series* (pp. 275–278). Elsevier.

Arias, C. A., & Murray, B. E. (2012). The rise of the *Enterococcus*: Beyond vancomycin resistance. *Nature Reviews Microbiology*, *10*(4), 266–278.

Asghar, M. T., Yusof, Y. A., Mokhtar, M. N., Yaacob, M. E., Ghazail, H. M., Chang, L. S., & Manaf, Y. N. (2020a). Coconut (*Cocos nucifera* L.) sap as a potential source of sugar: Antioxidant and nutritional properties. *Food Science & Nutrition*, *8*(4), 1777–1787. https://doi.org/10.1002/fsn3.1191

Asghar, M. T., Yusof, Y. A., Noriznan, M., Yaacob, M. E., Ghazali, H. M., Chang, L. S., & Manaf, Y. N. (2020b). Effect of processing method on vitamin profile, antioxidant properties and total phenolic content of coconut (*Cocos nucifera* L.) sugar syrup. *International Journal of Food Science & Technology*, *55*(7). https://doi.org/10.1111/ijfs.14529.

Asha, S., Ratheesh, M., Jose, S. P., Krishnakumar, I. M., & Sandya, S. (2019). Neera: A nonalcoholic nutritious beverage from unopened inflorescence of coconut palm. In *Natural Beverages* (pp. 339–360). Elsevier.

Barh, D., & Mazumdar, B. (2008). Comparative nutritive values of palm saps before and after their partial fermentation and effective use of wild date (*Phoenix sylvestris* Roxb.) sap in treatment of anemia. *Research Journal of Medicine and Medical Sciences*, *3*(2), 173–176.

Beegum, P. P. S., Nair, J. P., Manikantan, M. R., Pandiselvam, R., Shill, S., Neenu, S., & Hebbar, K. B. (2022). Effect of coconut milk, tender coconut and coconut sugar on the physico-chemical and sensory attributes in ice cream. *Journal of Food Science and Technology*. https://doi.org/10.1007/s13197-021-05279-y.

Carlini, N. A., Romanowski, S., Rabalais, E. N., Kistler, B. M., Campbell, M. S., Krishnakumar, I. M., Harber, M. P., & Fleenor, B. S. (2023). Coconut sugar derived from coconut inflorescence sap lowers systolic blood pressure and arterial stiffness in middle-aged and older adults: A pilot study. *Journal of Applied Physiology, 134*(3), 508–514. https://doi.org/10.1152/japplphysiol.00394.2022.

Chandrasekhar, K., Sreevani, S., Seshapani, P., & Pramodhakumari, J. (2012). A review on palm wine. *International Journal of Research in Biological Sciences, 2*(1), 33–38.

Chinnamma, M., Bhasker, S., Hari, B. M., Sreekumar, D., & Madhav, H. (2019). Coconut neera—a vital health beverage from coconut palms: Harvesting, processing, and quality analysis. *Beverages, 5*(1), 22. https://doi.org/10.3390/beverages5010022

Cortázar, R. M., Rogelio, F. F., & Fuentes del, A. I. M. (2010). Proceso productivo de la "tuba" de coco – una nueva alternativa económica para los cococultores del sureste mexicano. *Centro de Investigación Regional Sureste. Campo Experimental Chetumal*, 1, 43.

Devi, N. S., HariPrasad, T., Ramesh, K., & Merugu, R. (2015). Antioxidant properties of coconut sap and its sugars. *International Journal of PharmTech Research, 8*(1), 160–162.

Ezeagu, I. E., & Fafunso, M. A. (2003). Biochemical constituents of palm wine. *Ecology of Food and Nutrition, 42*, 213–222. https://doi.org/10.1080/03670240390226222

Faparusi, S. I. (1971). Microflora of fermenting palm wine. *Journal of Food Science and Technology, 8*, 206–210.

Filippini, T., Naska, A., Kasdagli, M. I., Torres, D., Lopes, C., Carvalho, C., Moreira, P., Malavolti, M., Orsini, N., Whelton, P. K., & Vinceti, M. (2020). Potassium intake and blood pressure: A dose-response meta-analysis of randomized controlled trials. *Journal of the American Heart Association, 9*, e015719. https://doi.org/10.1161/JAHA.119.015719

Gallegos, A. C. F., Vuelvas, O. F. V., Lopez, L. L. L., Galindo, A. S., Valdes, J. A. A., Aguilar, C. N., & Herrera, R. R. (2019). Tuba, a fermented and refreshing beverage from coconut palm sap. In *Non-alcoholic Beverages* (Vol. 6, pp. 163–184). The Science of Beverages. https://doi.org/10.1016/B978-0-12-815270-6.00006-2

Ghosh, D. K., Bandyopadhyay, A., Baidyanath, D., Kaushik, G., Hebbar, K. B., & Chowdappa, P. (2016). Method of the preparation of fresh coconut inflorescence sap (Kalparasa) based sweets. *Patent Application No. 201631013187*.

Gunawardhana, S. P. M. G., Hewapathirana, H. P. D. T., Yalegama, L. L. W. C., & Lakshman, P. L. N. (2018). Development of coconut sap based non-alcoholic ready to drink beverage. *Tropical Agricultural Research & Extension, 21*(1-2), 155–158. https://doi.org/10.4038/tare.v21i1-2.5463

Haddy, F. J., Vanhoutte, P. M., & Feletou, M. (2006). Role of potassium in regulating blood flow and blood pressure. *American Journal of Physiology: Regulatory, Integrative and Comparative Physiology, 290*, R546–R552. https://doi.org/10.1152/ajpregu.00491.2005

Hebbar, K. B., Arivalagan, M., Manikantan, M. R., Mathew, A. C., Thamban, C., Thomas, G. V., & Chowdappa, P. (2015). Coconut inflorescence sap and its value addition as sugar—Collection techniques, yield, properties, and market perspective. *Current Science, 109*(8), 1411–1417. https://doi.org/10.18520/v109/i8/1411-1417

Hebbar, K. B., Arivalagan, M., Manikantan, M. R., Mathew, A. C., Thamban, C., Thomas, G. V., & Chowdappa, P. (2015). Coconut inflorescence sap and its value addition as sugar—Collection techniques, yield, properties, and market perspective. *Current Science, 109*(8), 1411–1417. https://doi.org/10.18520/v109/i8/1411-1417

Hebbar, K. B., Arivalagan, M., Pavithra, K. C., Roy, T. K., Gopal, M., Shivashankara, K. S., & Chowdappa, P. (2020). Nutritional profiling of coconut (Cocos nucifera L.) inflorescence sap collected using novel coco-sap chiller method and its value-added products. *Journal of Food Measurement and Characterization, 14*, 2703–2712.

Hebbar, K. B., Ramesh, S. V., Ghosh, D. K., Beegum, P. P. S., Pandiselvam, R., Manikantan, M. R., & Mathew, A. C. (2022). Coconut sugar—a potential storehouse of nutritive metabolites, novel bio-products and prospects. *Sugar Tech, 24*(3), 841–856.

Ho, C. W., Aida, W. M. W., Maskat, M. Y., & Osman, H. (2007). Changes in volatile compounds of palm sap (Arenga pinnata) during the heating process for production of palm sugar. *Food Chemistry, 102*(4), 1156–1162.

Jayasundera, J., & Kulatunga, A. (2014). Spray-drying of coconut treacle into an amorphous powder. *Emirates Journal of Food and Agriculture, 26*(8), 672.

Jespersen, L. (2003). Occurrence and taxonomic characteristics of strains of *Saccharomyces cerevisiae* predominant in African indigenous fermented foods and beverages. *FEMS Yeast Research, 3*, 191–200. https://doi.org/10.1016/S1567-1356(02)00185-X

Jespersen, L. (2003). Occurrence and taxonomic characteristics of strains of *Saccharomyces cerevisiae* predominant in African indigenous fermented foods and beverages. *FEMS Yeast Research, 3*, 191–200. https://doi.org/10.1016/S1567-1356(02)00185-X

Jirasatid, S., & Nopharatana, M. (2020). Impact of coconut sugar syrup on quality of acidophilus milk during cold storage. *International Journal of Agricultural Technology, 16*(4), 819–830.

Jirovetz, L., Buchbauer, G., Fleischhacker, W., & Ngassoum, M. B. (2001). Analysis of the aroma compounds of two different palm wine species ('Matango' and 'Raffia') from Cameroon using SPME-GC-FID, SPME-GC-MS, and olfactometry. *Ernährung/Nutrition, 25*, 67–71.

John, S. Z., & Ushadevi, K. N. (2020). Socio-economic profile and purchase behaviour of elite consumers towards the coconut soft drink Neera. *International Research Journal of Agricultural Economics and Statistics, 11*(1), 42–48. https://doi.org/10.15740/HAS/IRJAES/11.1/42-48.

Jose, S. P., Asha, S., Krishnakumar, I. M., Ratheesh, M., Santhosh, S., Sandya, S., Kumar, G. B., & Pramod, C. (2017). Nephro-protective effect of a novel formulation of unopened coconut inflorescence sap powder on gentamicin-induced renal damage by modulating oxidative stress and inflammatory markers. *Biomedicine & Pharmacotherapy, 85*, 128–135. https://doi.org/10.1016/j.biopha.2016.11.117.

Joseph, A., Jose, S. P., Kalyan, B. T., Mammen, R. R., Krishnakumar, I. M., Fleenor, B. S., & Mohan, R. (2021). Coconut inflorescence sap enhances exercise performance and plasma antioxidant status in young active men. *NFS Journal, 23*, 37–43.

Kadere, T. T. (2021). Chemical composition and preservation of *mnazi* and its distillate (*Pyuwa*). *European Journal of Nutrition & Food Safety, 13*(1), 46–58. https://doi.org/10.9734/EJNFS/2021/v13i130347

Kadere, T. T., & Kutima, P. M. (2012). Isolation and identification of lactic acid bacteria in coconut toddy (*Mnazi*). *Journal of Asian Scientific Research, 2*(12), 807–819.

Kadere, T. T., Miyamoto, T., Oniang'o, R. K., Kutima, P. M., & Njoroge, S. M. (2008). Isolation and identification of the genera *Acetobacter* and *Gluconobacter* in coconut toddy (*mnazi*). *African Journal of Biotechnology, 7*(16), 2963–2971. http://www.academicjournals.org/AJB https://doi.org/10.5897/AJB08.390.

Kadere, T. T., Oniang'o, R. K., Kutima, P. M., & Njoroge, S. M. (2009). Marketing and economic importance of palm wine (*mnazi*) and other coconut-based products in Kenya. *Research Journal of Agricultural and Biological Sciences, 5*(5), 815–822.

Kadere, T. T., Oniango, R. K., Kutima, P. M., & Muhoho, S. N. (2004). Traditional tapping and distillation methods of coconut wine (*mnazi*) as practised in the coastal region of Kenya. *African Journal of Food, Agriculture, Nutrition and Development, 4*(1), 1–16. https://doi.org/10.4314/ajfand.v4i1.19153.

Kalaiyarasi, K., Sangeetha, K., & Rajarajan, S. (2013). A comparative study on the microbial flora of the fresh sap from cut inflorescence and fermented sap (toddy) of *Borrassus flabellifer* Linn (Palmyrah tree) and of *Cocos nucifera* Linn (Coconut tree) to identify the microbial fermenters. *International Journal of Research in Pure and Applied Microbiology, 3*(3), 43–47.

Karamoko, D., Djeni, N. T., N'guessan, K. F., Bouatenin, K. M. J. P., & Dje, K. M. (2012). The biochemical and microbiological quality of palm wine samples produced at different periods during tapping and changes which occurred during their storage. *Food Control, 26*, 504–511.

Karseno, Erminawati, Yanto, T., & Handayani, I. (2021). The effect of coconut sap and skim milk concentration on physicochemical and sensory characteristics of coconut sap drink yogurt. *IOP Conference Series: Earth and Environmental Science, 653*, 012045. https://doi.org/10.1088/1755-1315/653/1/012045.

Kaur, G., & Goswami, T. K. (2021). Physical and sensory characteristics of low sugar dairy dessert (rasgulla) developed at different levels of coconut sap syrup. *Journal of Food Science and Technology, 58*(1), 343–348.

Kochukalam, C. (2014). A coconut health drink for the revival of the coconut farming industry. Available at SSRN: https://ssrn.com/abstract=2717268.

Kusumawaty, Y., Maharani, E., & Edwina, S. (2012). Perceived quality of coconut sugar by producers, traders, and downstream industries in Indragiri Hilir District, Riau Province, Indonesia. *Journal of Agribusiness Marketing, 5*, 1–13.

Mahulette, F., Mubarik, N. R., Suwanto, A., & Widanarni. (2016). Isolation and characterization of lactic acid bacteria from Inasua. *Journal of Tropical Biodiversity and Biotechnology*, 1(2), 71–76.

Mammen, R., John, F., Krishnakumar, I. M., Maliakel, B., Mohanan, R., & Fleenor, B. (2020). Safety and influence of a novel powder form of coconut inflorescence sap on glycemic index and lipid profile. *Bioactive Carbohydrates and Dietary Fibre, 23*, 100217. https://doi.org/10.1016/j.bcdf.2020.100217.

Manikantan, M. R., Beegum, P. P. S., Pandiselvam, R., & Hebbar, K. B. (2020). Coconut milk Peda. *Kalpa CPCRI Newsletter, 39*(1), 4.

Martins, S. I. F. S., Jongen, W. M. F., & Van Boekel, M. A. J. S. (2000). A review of Maillard reaction in food and implications to kinetic modelling. *Trends in Food Science & Technology, 11*(9–10), 364–373.

Melgar, A. F., Leyva, O. A., Utrilla, J., & Beristain, H. G. (2019). Bacterial diversity and population dynamics during the fermentation of palm wine from Guerrero, Mexico. *Frontiers in Microbiology, 10*, 531. https://doi.org/10.3389/fmicb.2019.00531.

Mesquita, M. C., Leandro, E. D. S., de Alencar, E. R., & Botelho, R. B. A. (2020). Fermentation of chickpea (*Cicer arietinum* L.) and coconut (*Cocos nucifera* L.) beverages by *Lactobacillus paracasei* subsp. *paracasei* LBC 81: The influence of sugar content on growth and stability during storage. *LWT, 132*, 109834. https://doi.org/10.1016/j.lwt.2020.109834.

Misra, B. (2016). Neera: The coconut sap: A review. *International Journal of Food Science and Nutrition, 1*(4), 35–38.

Mithra, A., Swamy, G. J., Chandrasekar, V., & Shanmugam, S. (2013). Coconut value-added products. *Journal of Indian Food Industry Magazine, 32*(6), 29–36.

Monreal, V. J., Rentería, M. N. P., Herrera, R. R., & González, A. C. N. (2011). Tuba: Una bebida fermentada tradicional de Colima. *Cienciacierta, 25*, 11–14.

Mummaleti, G., Sarma, C., Kalakandan, S. K., Gazula, H., Sivanandham, V., & Anandharaj, A. (2022). Characterization of levan produced from coconut inflorescence sap using *Bacillus subtilis* and its application as a sweetener. *LWT*, 112697. https://doi.org/10.1016/j.lwt.2021.112697.

Mummaleti, G., Sarma, C., Kalakandan, S., Sivanandham, V., Rawson, A., & Anandharaj, A. (2021). Optimization and extraction of edible microbial polysaccharide from fresh coconut inflorescence sap: An alternative substrate. *LWT*, 138, 110619. https://doi.org/10.1016/j.lwt.2020.110619.

Muralidharan, K., & Jayashree, A. (2011). Value addition, product diversification and by-product utilization in coconut. *Indian Coconut Journal*, 7, 4–10.

Murugesan, P., Moses, J. A., & Chinnaswamy, A. (2020). One-step synthesis of fluorescent carbon dots from neera for the detection of silver ions. *Spectroscopy Letters*, 53(6), 1–9. https://doi.org/10.1080/00387010.2020.1764589.

Naik, A., Madhusudhan, C. M., Raghavarao, K., & Subba, D. (2015). Downstream processing for production of value-added products from coconut. *Current Biochemical Engineering*, 2(2), 168–180.

Njugu, S. P. (2013). Competitiveness of coconut wine value chain in the coconut industry in Kenya. *Developing Country Studies*, 3(6), 108–114.

Nurhadi, B., Sukri, N., & Sugandi, W. K. (2018). Comparison of crystallized coconut sugar produced by traditional method and amorphous coconut sugar formed by two drying methods: Vacuum drying and spray drying. *International Journal of Food Properties*, 21(1), 2339–2354. https://doi.org/10.1080/10942912.2018.1517781.

Okafar, N. (1975). Microbiology of Nigerian palm wine with particular reference to bacteria. *Journal of Applied Bacteriology*, 38(2), 81–88. https://doi.org/10.1111/j.1365-2672.1975.tb00507.x.

Okoma, D. M. J., Konan, K. J. L., & Assa, R. R. (2020). Vitamin and osidic composition of table sugars from the inflorescence sap of three coconut cultivars (*Cocos nucifera* L.) in Ivory Coast. *Food and Nutrition Sciences*, 11(12), 1117–1126.

Pandiselvam, R., Akshay, A. T., Manikantan, M. R., Ramesh, S. V., Patil, H., & Gopal, M. (2023). Synergetic effect of fermented coconut inflorescence sap for the production of virgin coconut oil. *Journal of Food Biochemistry*, 2023, 7125810.

Pravitha, M., Manikantan, M. R., Ajesh Kumar, V., Beegum, S. P. P., & Pandiselvam, R. (2021). Optimization of process parameters for the production of jaggery-infused osmo-dehydrated coconut chips. *LWT*, 146(1), 111441. https://doi.org/10.1016/j.lwt.2021.111441.

Purnomo, H. (2007). Volatile components of coconut fresh sap, sap syrup, and coconut sugar. *ASEAN Food Journal*, 14(1), 45–49.

Rajath, K., & Santhoshkumar, N. (2020). Effectiveness of marketing strategy of 'neera health drink' with reference to the 4P's of marketing. *International Journal of 360 Management Review*, 7(4), 1–14.

Rajesh, M. K., Muralikrishna, K. S., Nair, S. S., Krishna, K. B., Subrahmanya, T. M., Sonu, K. P., Subaharan, K., Sweta, H., Keshava, P. T., Neeli, C., & Karunasagar, I. (2020). Facile coconut inflorescence sap mediated synthesis of silver nanoparticles and its diverse antimicrobial and cytotoxic properties. *Materials Science and Engineering: C*, 111, 110834.

Ratheesh, M., Svenia, J. P., Asha, S., Sandya, S., Girishkumar, B., & Krishnakumar, I. M. (2017). Anti-inflammatory effect of a novel formulation of coconut inflorescence sap against ox-LDL induced inflammatory responses in human peripheral blood mononuclear cells by modulating TLR-NF-κB signaling pathway. *Toxicology Mechanisms and Methods*, 27, 615–621. https://doi.org/10.1080/15376516.2017.1344339.

Ravindran, V., Sriskandarajah, N., & Rajaguru, A. S. B. (1984). The utilization of coconut sap distillery by-products by broiler chicks. *Agricultural Wastes*, 11(2), 115–124.

Renjith, R. S., Chikku, A. M., & Rajamohan, T. (2013). Cytoprotective, antihyperglycemic, and phytochemical properties of *Cocos nucifera* (L.) inflorescence. *Asian Pacific Journal of Tropical Medicine*, 6(10), 804–810. https://doi.org/10.1016/S1995-7645(13)60142-X.

Rossman, M. J., Ryan, G. R. A., Parker, S. J. R., Ziemba, B. P., Lubieniecki, K. L., Johnson, L. C., Poliektov, N. E., Bispham, N. Z., Woodward, K. A., Nagy, E. E., Bryan, N. S., Reisz, J. A., D'Alessandro, A., Chonchol, M., Sindler, A. L., & Seals, D. R. (2021). Inorganic nitrite supplementation improves endothelial function with aging: Translational evidence for suppression of mitochondria-derived oxidative stress. *Hypertension*, 77, 1212–1222. https://doi.org/10.1161/Hypertensionaha.120.16175.

Saidan, N. H., Kartayasa, M. A., Lokman, N. I., Ab Llah, N. F., Nor, M. M., Mokhtar, S. I., Zaman, K. A., Sidek, N. M., & Hamil, M. S. (2020). Effect of chengal wood (*Neobalanocarpus heimii*) on physicochemical and antimicrobial properties of coconut sap (*Cocos nucifera*). *IOP Conference Series: Earth and Environmental Science*, 549, 012073. https://doi.org/10.1088/1755-1315/549/1/012073.

Salcido, N. M. F., Ramírez, J. C. C., Almendárez, G. E. G., Bideshi, D. K., Hernández, R. S., & Corona, J. E. B. (2015). Isolation and characterization of bacteriocinogenic lactic bacteria from M-Tuba and tepache, two traditional fermented beverages in México. *Food Science & Nutrition*, 3(5), 434–442. https://doi.org/10.1002/fsn3.236.

Sánchez, D. G., & Ríos, G. F. L. (2002). Manejo de la palma de coco (*Cocos nucifera* L.) en México (Management of the coconut palm (*Cocos nucifera* L.) in Mexico). *Revista Chapingo Serie Ciencias Forestales y del Ambiente*, 8(1), 39–48.

Sanchez, P. C. (2009). Philippine fermented foods: Principles and technology (Chapter 8, pp. 151–173). The University of the Philippines Press. 516 pages.

Santoso, H. B. (1993). *Pembuatan Gula Kelapa (Making coconut sugar)*. Penerbit Kanisius: Yogyakarta. 67 pages.

Sarma, C., Mummaleti, G., Sivanandham, V., Kalakandan, S., & Anandharaj, A. (2022). Anthology of palm sap: The global status, nutritional composition, health benefits, and value-added products. *Trends in Food Science & Technology*, 119, 530–549. https://doi.org/10.1016/j.tifs.2021.12.002.

Sharma, A., & Meghwal, M. (2021). An update on uses, benefits, and potential application of Neera. *Journal of Nutrition and Food Processing*, 4(8). https://doi.org/10.31579/2637-8914/072.

Sharma, R. K., Chauhan, O. P., & Xavier, J. R. (2021). Technological innovations in food processing and value addition to coconut. Journal of Food and Agricultural Research, 1, 69–85.

Singaravadivel, K., Alagusundaram, K., & Hariharan, B. (2012). Physicochemical properties of fresh and stored coconut palm toddy. *Open Access Scientific Reports*, 1(8), 2 pages. https://doi.org/10.4172/scientificreports.397.

Singhal, R. S., Kennedy, J. F., Gopalakrishnan, S. M., Kaczmarek, A., Knill, C. J., & Akmar, P. F. (2008). Industrial production, processing, and utilization of sago palm-derived products. *Carbohydrate Polymers*, 72(1), 1–20.

Sohn, M., & Ho, C. T. (1995). Ammonia generation during thermal degradation of amino acids. *Journal of Agricultural and Food Chemistry*, 43(12), 3001–3003.

Somashekaraiah, R., Shruthi, B., Deepthi, B. V., & Sreenivasa, M. Y. (2019). Probiotic properties of lactic acid bacteria isolated from neera: A naturally fermenting coconut palm nectar. *Frontiers in Microbiology*, 10, 1382.

Srikaeo, K., & Thongta, R. (2015). Effects of sugarcane, palm sugar, coconut sugar, and sorbitol on starch digestibility and physicochemical properties of wheat-based foods. *International Food Research Journal*, 22(3), 923–929.

Sudana, I. M., & Waisnawa, I. G. N. S. (2020). The influence of coconut sap heating temperature on the content of alcohol elements in arak distillation process. *Journal of Physics: Conference Series*, 1450, 012104. https://doi.org/10.1088/1742-6596/1450/1/012104.

Swamy, G. M. S. (2013). Coconut neera production and processing in Karnataka. *Indian Coconut Journal*, 56, 31–33.

Udomsaksakul, N., Kodama, K., Tanasupawat, S., & Savarajara, A. (2018a). Diversity of ethanol fermenting yeasts in coconut inflorescence sap and their application potential.

ScienceAsia, 44, 371–381. https://doi.org/10.2306/scienceasia1513-1874.2018.44.371.

Udomsaksakul, N., Kodama, K., Tanasupawat, S., & Savarajara, A. (2018b). Indigenous *Saccharomyces cerevisiae* strains from coconut inflorescence sap: Characterization and use in coconut wine fermentation. *CMU Journal of Natural Sciences*, 17(3), 219–230. https://doi.org/10.12982/CMUJNS.2018.0016.

Vaulina, E., Widyaningsih, A., Kartika, D., & Romdoni, M. P. (2018). The effect of cellulose acetate concentration from coconut nira on ultrafiltration membrane characters. *IOP Conference Series: Materials Science and Engineering*, 349, 012020. https://doi.org/10.1088/1757-899X/349/1/012020.

Velasco, R. M. (2013). Gender responsive value chain analysis of the Lambanog industry in the Philippines. *International Journal of Business and Economic Development*, 2, 41–53.

Wattimena, S. C., Temartenan, J. S., & Lesbatta, K. J. (2021). The effect of coconut sap amount and salt texture on the protein content and total bacterial number in *ina sua*. *Journal of Physics: Conference Series*, 1943(1), 012064.

Widyaningsih, S., Chasani, M., Zusfahair, Anam, M. S. K., & Anisya, A. N. (2020). The effect of additive on cellulose acetate membrane from coconut sap in desalination application. *Malaysian Journal of Science*, 39(2), 41–50.

Wu, F. C., Chou, S. Z., & Shih, I. L. (2013). Factors affecting the production and molecular weight of levan of *Bacillus subtilis* natto in batch and fed-batch culture in fermenter. *Journal of the Taiwan Institute of Chemical Engineers*, 44(6), 846–853. https://doi.org/10.1016/j.jtice.2013.03.009.

Yoghapriya, R., & Sivakami, P. L. S. (2011). Formulation and physicochemical analysis of neera banam powder. *Journal of Huazhong University of Science and Technology*, 50(4), 1–12.

Ysidor, K. N. G., Konan, K. J. L., Roger, K. B., Assa, R. R., Joelle, O. D. M., Emmanuel, I. A., & Marius, B. G. H. (2015). Changes in physicochemical parameters during storage of the inflorescence sap derived from four coconut (*Cocos nucifera* L.) varieties in Côte d'Ivoire. *American Journal of Experimental Agriculture*, 5(4), 352–365.

Yuvarani, R., Shoba, N., Senthamizh Selvi, B., & Vennila, P. (2020). Studies on development of neera powder from coconut (*Cocos nucifera* L.) neera by using spray dryer. *Journal of Pharmacognosy and Phytochemistry*, 9(6), 2135–2138.

Chapter 8

The Nutritional Profiling of Coconut Sugar and Other Derived Products

S. V. Ramesh[1]
V. Krishnakumar[2]
J. Anjitha[1]
S. Sudharshana[1]
and K. B. Hebbar[1,*]
[1]ICAR-Central Plantation Crops Research Institute, Kasaragod, Kerala, India
[2]ICAR-Central Plantation Crops Research Institute, Kayamkulam, Kerala, India

Abstract

Coconut sugar is considered a healthy sweetener compared to most other commercially available sugars. With the increasing public interest in healthy diets and the negative perception associated with the excessive sugar consumption, consumers often seek to replace refined sugars with alternative sweeteners such as coconut sugar. The coconut sugar production process involves collecting inflorescence sap from palms, which is then transported, stored, and evaporated. These steps are labour- and resource-intensive, resulting in higher production costs compared to production of cane sugar. This chapter explores into the nutritional profile and health benefits of coconut sugar positioning it as a natural sweetener of significant interest. It also examines the various applications of coconut sugar in the food industry, addressing sustainability concerns and production methods. Additionally, the chapter highlights quality criteria and explores an array of value-added products derived from

* Corresponding Author's Email: hebbar.kb@icar.gov.in

In: Coconut Palm Sap
Editors: K. B. Hebbar, V. Krishnakumar and S. V. Ramesh
ISBN: 979-8-89530-640-6

coconut sugar, emphasizing its potential to meet the growing consumer demand for natural ingredients.

Keywords: nutritional profile, glycemic index, coconut, palm sap, browning intensity

1. Introduction

The coconut tree (*Cocos nucifera* L.) holds significant cultural and practical importance as one of the most utilized palm species globally. Often referred to as the 'tree of life' or '*Kalpavriksha,*' it has been cultivated across various tropical regions worldwide for its diverse uses and benefits. The rapid growth of the coconut and coconut-based products market can be attributed to its nutritional and health promotional properties (Henrietta et al., 2022).

Source: ICAR-CPCRI.

Figure 1. Coconut sugar.

Coconut inflorescence sap, known as *neera* in India, *nira* in Indonesia, *maprau* in Thailand, and *lagbi* in North Africa, serves as a primary ingredient in the manufacturing process of coconut sugar. Coconut sugar (Figure 1), known for its solid, dry, and brownish-yellow appearance, derives its traits from the sap or juice used for its production. Optimal results are achieved when the pH of inflorescence sap ranges between 6.5 to 7.0 before heating. This pH range ensures the development of desirable pigments and the distinctive flavour characteristic of coconut sugar.

Today's consumers are becoming more conscious of the health benefits of natural ingredients, which has greatly influenced the food industry (Roman et al., 2017; Saraiva et al., 2020). This shift has led to a dramatic increase in the demand for sweeteners derived from natural sources in recent decades (Philippe et al., 2014). The demand for coconut palm sugar is on the rise due to an increasing interest in low-calorie, reduced-sugar, and sugar-free products, as well as greater health awareness in food markets. This trend is partly driven by the growing number of individuals suffering from dental issues, obesity, and diabetes (Hebbar et al., 2015). Coconut sugar is used as a sweetener in various food and beverage products, including chocolate, baked goods, snacks, and breakfast cereals. It is often promoted as a more environmentally friendly alternative to cane sugar, gaining acceptance among eco-conscious consumers (Witham 2016).

Studies on coconut sap sugar and syrup revealed that both the products have a low glycemic index (GI) of 35 and 39, respectively. Foods with a low GI are beneficial for health as they help control the increase in blood glucose levels, which can play a significant role in reducing the risk of diabetes mellitus (DM) and obesity. This discovery marks a significant advancement in the sugar industry, providing a naturally sweet alternative for individuals suffering from DM. The considerable amounts of potassium, vitamin C, inulin, and propionate found in coconut sap sugar/syrup add value to these products, making them promising as functional foods/ingredients. The potassium and propionate produced from inulin fermentation in both coconut sap sugar and syrup could potentially play a role in preventing the risk of coronary heart disease. Additionally, the inulin and dietary fiber content of coconut sap sugar and syrup, respectively, might contribute to the slow release of glucose after consumption, thereby helping in managing and controlling DM.

According to Coconut Sugar Market Research, the global coconut sugar market was valued at $243.4 million in 2021 and is expected to grow to $408.7 million by 2031, with a compound annual growth rate (CAGR) of 5.5% from 2022 to 2031. The Philippines, Indonesia, and Thailand are the leading producers, accounting for approximately 80% of global production over the past few decades (source: https://www.alliedmarketresearch.com/coconut-sugar-market-A16949). Coconut sugar is versatile due to its lower calorie content and GI compared to regular sugar, making it popular in beverages, confections, and pastries. Its unique properties and increased availability make coconut sugar an appealing option for food manufacturers and health conscious consumers.

Given this context, this chapter explores the nutritional profile and health impacts of consuming coconut sugar. It also examines its applications in the food industry, sustainability issues, production methods, quality standards, and various value-added products.

2. Nutritional Profile of Coconut Sugar

Coconut sugar is a traditional sweetener in Asian countries and is globally recognized as a 'nutritious' and 'low glycemic' natural sweetener. Food attributes such as colour, consistency, texture, flavour, and aroma are crucial for consumers and, by extension, for the industry. In the case of coconut sugar and syrup, these characteristics are influenced by the quality of the sap used in their production and the chemical reactions that take place during the heating process. The nutritional composition of coconut sugar is listed in Table 1.

Specifically, non-enzymatic browning, including caramelization and Maillard reactions plays a significant role in developing these attributes. Coconut sugar and syrup consist of over 100 distinct compounds, including carbohydrates, proteins, minerals, free amino acids, vitamins, phenolics, and aromatic compounds. The primary amino acids in fresh coconut *neera* are glutamic acid, threonine, aspartic acid, and serine, while proline, methionine, tryptophan, and histidine are present in quantities of less than 4.00 mg $100g^{-1}$ (Purnomo 2007). The physical and chemical composition of coconut sap indicates a high content of amino acids, particularly glutamic acid. The amino acid content (g $100g^{-1}$) of freshly gathered coconut sap includes histidine at 1.19, arginine at 0.35, aspartic acid at 11.22, threonine at 15.36, serine at 8.24, glutamic acid at 34.20, proline at 3.52, glycine at 0.47, alanine at 2.5, valine at 2.11, isoleucine at 0.48, tyrosine at 0.31, and phenylalanine at 0.78. This indicates that coconut sap is indeed rich in amino acids, particularly glutamic acid.

Coconut sugar, with a sweetening potential similar to saccharose, is used as an alternative sweetener in the preparation of confections, beverages, pastries, and various other culinary delights. It is nutritionally superior compared to other sugars. When mixed with wheat flour, sweeteners such as cane, palm, and coconut sugars, as well as sorbitol, are evaluated for starch digestibility. Sorbitol exhibits the highest starch digestibility, with palm and coconut sugars showing nearly identical digestibility levels (Srikaeo and Thongta 2015).

Nakamura et al., (2004) identified the composition of coconut sugar derived from sap, as it contains 339 mM of sucrose, a total amino acid concentration of 17.1 mM, 48.3 mM of potassium, and a protein concentration of 0.1 $\mu g\ \mu l^{-1}$ in the exudate. According to Trinidad (2015), coconut sap sugar maintains acceptable water activity over time. Its moisture content, ranging from 1.7 to 1.8 g 100 g^{-1}, complies with CODEX Standards. Additionally, levels of Salmonella and Coliform were found to be within the acceptable limits set by the Philippine National Standards. Although coconut sap sugar does not contain dietary fiber, it has a notable amount of inulin at 4.7 g 100 g^{-1}. This significant presence of inulin and propionate adds nutritional value to coconut sap sugar, positioning it as a promising functional food or ingredient.

It contains nearly 80% sucrose, with 3-7% glucose and fructose. Studies have reported the glycemic index (GI) of coconut sugar to be 35 and 54, in contrast to the GI of table sugar, which is 70. While the fresh sap from coconut inflorescences is reported to contain vitamins (B1, B3, B5, B7, B9, and C) and minerals (Na, K, Ca, Mg, P, Fe, and Zn), the majority of these nutrients are lost during the evaporation process used to produce coconut sugar (Jose et al., 2017).

Table 1. Nutritional composition of coconut sugar (Hebbar et al., 2020)

Biochemical Constituents	Sap sugar (100 g)
Total sugars (g)	91.8 ± 1.01
Non-reducing sugars (g)	87.3 ± 5.91
Reducing sugars (g)	4.69 ± 4.60
Free amino acids (g)	3.05 ± 0.07
Total phenolics (mg GAE)	47.2 ± 20.7
Total flavonoids (mg CE)	4.76 ± 1.21
FRAP (mg AEAC)	22.9 ± 4.12

GAE: Gallic Acid Equivalent; CE: Catechin Equivalent; AEAC: Ascorbic Acid Antioxidant Capacity

Coconut sugar typically contains 70-80% sucrose, while jaggery has about 50% sucrose. The caloric content of coconut sugar is similar to that of table sugar, providing 16 calories per teaspoon. Due to minimal processing, coconut sugar retains various natural vitamins, minerals (notably iron, zinc, calcium, and potassium), inulin (a type of fiber), and antioxidants. Research by Kurniawati et al., (2019) indicated that coconut palm sap sugar produced in Indonesia had a moisture content of 2.55%, total ash of 1.81%, crude fat of 0.6%, crude protein of 0.28%, and total solids of 48.6%, with a GI of 53. Organoleptic testing revealed quality scores for color, taste, aroma, and texture of 4.23, 4.25, 3.88, and 4.0, respectively, indicating consumer acceptance.

Putra et al., (2019) explored the fermentation of coconut sap using 4.5% guava leaves, corn cobs, and mangosteen peels, resulting in sugar with a moisture content of 0.24%, ash content of 1.3%, reducing sugar content of 0.39%, and sucrose level of 69.99%. The resulting coconut sugar was brownish-yellow, with a sweet flavor and typical coconut sugar aroma.

Coconut palm sap treated with mangosteen peel at a 4.5% concentration produced sugar that met Indonesian National Standard (SNI) requirements, with moisture, ash, reducing sugar, and total sugar contents of 7.50%, 2.0%, 7.30%, and 83.94%, respectively (Naufalin et al., 2013). Haloho and Susanto (2015) treated fresh inflorescence sap with 22% lime and 400 ppm sodium tripolyphosphate, yielding coconut sugar with a moisture content of 4.42%, ash content of 3.46%, reducing sugar of 4.22%, and total sugar of 79.62%. Another study indicated that coconut sugar produced from sap treated with 0.84 g l^{-1} mangosteen peel powder exhibited good chemical properties, including reducing sugar (5.01 g100 g^{-1}), total sugar (86.50 g100 g^{-1}), total phenolic content (470 mg GAE 100 g^{-1}), browning index (16254 IU), and antioxidant activity (61.05% RSA) (Haryanti et al., 2018).

Zulfia et al., (2019) determined the optimal conditions of preparation of coconut sugar using a combination of 0.1% w/v mangosteen peel, 0.07% w/v crushed wood powder, and 2% v/v sodium metabisulfite, along with lime solution as sap preservatives, coconut milk as a defoaming agent, an 8 h morning and 16 h afternoon tapping duration, and cooking at 135°C. This method produced sugar with 6.87 ± 0.57% moisture, 1.96 ± 0.13% ash, 77.77 ± 5.19% sucrose, and 8.09 ± 4.28% reducing sugar. Zulfia et al., (2022) further confirmed that combining mangosteen peel and lime as sap preservatives, using coconut milk as a defoaming agent, and maintaining a long tapping time of 8 h plus 16 h at 135°C yielded a 15.58% production efficiency.

Kurniawati et al., (2021) found that adding sodium metabisulfite raised the pH, impacting reducing sugar levels and the crystallization process. The best results in coconut sugar production occurred when 0.2% sodium metabisulfite was used and a drying time of 15 min. was maintained, leading to a product characterised with 2.41% moisture, 1.77% total ash, 3.34% reducing sugar, and an 82% yield.

Comparative studies of various sugars (cane, palm, and coconut) regarding physicochemical, antioxidant, and volatile flavour properties revealed that unrefined coconut sugar (sourced from Thailand) had a significantly lower ICUMSA colour value than unrefined cane sugar while maintaining comparable total soluble solid (TSS) levels (24°Brix) (Weerawatanakorn et al., 2021). In contrast to other sugars, coconut sugar had the lowest levels of phenolic compounds (0.6 to 3.3 mg GAE 100 g^{-1}), total flavonoids (9.2 to 14.0 mg RE 100 g^{-1}), DPPH radical scavenging activity (2.05 to 2.25% inhibition), and oxygen radical absorbance capacity (ORAC) (30.5 to 31.2 mg TE 100 g^{-1}).

The reducing sugar and amino acid content are critical chemical characteristics of coconut sap that influence the colour and flavour of resultant coconut sugar. Haryanti and Sulistyo (2022) noted that adding lysine to sap did not significantly affect the physico-chemical properties (moisture, ash, total sugar, reducing sugar, sucrose, free amino acids, and total phenolics) of granulated coconut sugar. However, the amount of lysine added significantly impacted the browning intensity, radical scavenging, and chelating activities. The highest radical scavenging activity of 73.96% was recorded with 0.5 mM lysine, while 0.25 mM lysine resulted in the highest chelating activity at 7.01%.

Hanifah et al., (2022) investigated the physical characteristics of palm sugar, revealing water content between 1.55% and 2.43%, sugar yield ranging from 13.57% to 24.37%, and particle sizes of 1.21 to 1.39 mm. Zulfia et al., (2022) attempted to optimize coconut sugar production parameters using the Taguchi design in Riau, Indonesia. This method considered four variables: sap preservative type, defoaming agent, tapping time, and cooking temperature. It had led to the identification of optimal parameters as the combination of mangosteen peel and lime as preservatives, coconut milk as a defoaming agent, an 8 h morning and 16 h tapping duration, and a cooking temperature of 135°C, resulting in a yield of 15.58% with moisture content of 6.87%, ash content of 1.96%, sucrose content of 77.77%, and reducing sugar content of 8.09%, with organoleptic test scores close to 4 (indicating likability) and compliance with SNI standards.

Mineral content (mg 100 g^{-1}) in coconut sugar samples ranged from 101.77 to 128.95 for potassium, 85.32 to 94.66 for chloride (Cl), 7.96 to 16.28

for magnesium, 12.68 to 15.87 for silicon, 8.33 to 14.57 for phosphorus, 5.58 to 13.17 for sulfur, 8.05 to 11.65 for sodium, 1.23 to 2.19 for copper, and 1.73 to 2.09 for iron, with traces of bromine at 1.04 and zinc at 0.17 (Muriel et al., 2021).

Coconut sugar contains vitamins C and E, minerals such as zinc, iron, potassium, and phosphorus, as well as phytonutrients like antioxidants, flavonoids, anthocyanidins, and polyphenols. Furthermore, coconut sap sugar contains a significant amount of inulin (4.7 g 100 g^{-1}). This inulin is essential for the production of short-chain fatty acids such as acetate, butyrate, and propionate. Phenolic compounds with antioxidant properties play a crucial role in reducing oxidative stress, making them desirable constituents of human diets. Coconut sugar is a rich source of dietary phenolics. Coconut sugar contains all four minerals (Cu, Fe, Zn, and Mn). Palm sugar contains Fe, Zn, and Cu, whereas refined sugar cane contains only Fe and Zn. Fluctuations in nutrient content are common and can be influenced by factors such as plant varieties, cultivation practices, soil, climate, and cultural practices (Luchini et al., 2017). Small quantities of mineral elements are essential for maintaining good health, but an excess of these elements can lead to acute or chronic toxicity. The high iron content in coconut sugar is beneficial as it prevents anaemia.

Coconut sugar stands out as a healthier sweetener compared to most commercially available sugars. Its production involves labour-intensive and resource-demanding processes, starting with the collection of sap from trees. This sap is then transported, stored, and evaporated to produce the sugar. Consequently, the manufacturing cost is higher than that of cane sugar. However, due to its nutritional benefits and low glycemic index, consumers are willing to pay a premium price for it.

Water activity ($_a$w) refers to the level of water in a product that is accessible for microorganism growth. In accordance with the Philippine National Standard for grading coconut sap sugar, the water activity ($_a$w) of this sugar typically falls between 0.5 and 0.8. However, the United Nations' Food and Agriculture Organization has designated coconut sugar as the 'most sustainable sweetener in the world.' This is because coconut trees require significantly less water than sugarcane and can produce yields for over 25 years (Mammen et al., 2020).

Based on the odour analysis conducted by Kabir and Lorjaroenphon (2014), aroma compounds in coconut sugar were classified into five categories. The identified compounds include pyrazine, furan, pyran derivatives, ketones, aldehydes, alcohols, acids, and sulfur-containing compounds. The majority of the aroma compounds were heterocyclic in nature, which included: 2,5-dimethyl pyrazine (nutty, sweet, roasty); Furfural (smoky, burnt, bread-like, caramel); 5-methyl furfural (strong burnt sugar, nutty, sweet); 2-furan methanol (caramel, burnt sugar, musty, sweet); 4-methyl-5H-furan-2-one (roast, nutty, sweet, caramel); 5-methyl-2-furan methanol (sweet, caramel); Maltol (sweet, caramel, cotton candy); Furaneol® (burnt sugar, caramel, cotton candy). These heterocyclic compounds contribute to the sweet, burnt, roasty, nutty, smoky, and caramel characteristics of coconut sugar. Additionally, ketones were associated with buttery, cheesy, and creamy notes. Benzyl alcohol and vanillin added sweet aromas, while acetic acid, known for its sharp, pungent note, represented a separate group. Methional, which has a French fry's aroma, was categorized in the fourth group. The miscellaneous group included two unidentified compounds: Unknown-3 (which has an ethereal note) and Unknown-7 (which has a fenugreek-like scent). Furthermore, Purnomo (2007) identified various compounds in coconut sugar, including: acetic acid (21.54–35.05%); 2-butanol (29.98–31.23%); 1,4-dimethyl-6,1-butyl acetate (1.7–15.50%); N,N-dimethyl-2-(diphenylmethoxy)-ethylamine (9.31–13.26%); Cyclohexyl-octane (0.0–17.01%); dodecanoic acid (0.0–12.41%); methylpyrazine (1.46–1.81%).

3. Method of Production of Coconut Palm Sugar

Coconut palm sugar serves as natural sweetener with distinctive taste and colour which cannot be easily substituted. It also serves a natural brown colour agent for large number of foods and beverages (Kroger et al., 2006; Saraiva et al., 2020) produced using the inflorescence sap of coconut palm (Indrayanto et al., 2022; Mammen et al., 2020). When the sap is unfermented and retains its original form, it contains 14% sugar. As fermentation begins, the sugar content decreases and it crystallizes easily upon heating without any chemical additives. Sap with a pH greater than 6.7 is ideal for producing sugars (Hebbar et al., 2015). Boiling the sap with lower pH results in either a semi-solid gum-like structure or hard lumps. The freshly harvested, coconut inflorescence sap was heated in a specially designed double-walled cooker at 115°C to evaporate the water content (Hebbar et al., 2022). During heating, the sugars in the palm sap undergo the Maillard reaction and caramelization. The volatile components of coconut and palm sugar, such as pyrazines, furans, ketones, fatty acids, and organic acids, contribute to their sweet, toasty, and nutty caramel-like aroma (Sohn and Ho 1995). The thick sap is continuously heated and stirred until crystals form, then rapidly cooled to produce sugary granules. Continuous stirring during cooling prevents clump formation and subsequent sieving ensures the creation of uniform-sized coconut palm sugar granules. Different processing methods are based on the type of heating and moisture extraction from the sap.

3.1. Conventional Method

Traditionally, coconut sap is heated at high temperatures (100–120°C) for several hours (Asghar et al., 2021; Prakobsil et al., 2010) in an open pan, a method that often leads to over-cooking of sugar and degradation of its quality (Apriyantono et al., 2002). Browning reactions occur at higher temperatures and extended processing times, further impacting the sugar's quality (Karseno et al., 2018). Such high temperature heating also results in a reduction of vitamin C and vitamin B3 content, diminishing the overall quality of coconut sugar (Asghar et al., 2020a).

Ho et al., (2007) described the sugar production process, where a substantial amount of filtered sap is placed in a large wok and heated on a wood or gas-fired stove for 3 to 5 h at 100°C until it reaches a concentrated state, accompanied by a distinctive aroma. During this heating process, two primary reactions occur: the Maillard reaction and caramelization. Once the sap is concentrated, it is poured into steel or bamboo moulds to create solid sugar, as detailed by Ho et al., (2008).

The crystallization of sugar becomes challenging in partially fermented *neera* with lower sugar content. To address this issue, various additives such as gluten, starch, sugars from C4 plants, coconut, or palm oil are commonly used during heating in many regions. Coconut quarters are sometimes added during the sugar preparation process to prevent over-boiling, while oil helps prevent foam formation, reducing sugar yield loss. Starch, often derived from wheat, is added to improve pourability and prevent agglomeration by absorbing excess moisture. Powdery products and plant extracts, acting as release agents, are also incorporated to absorb moisture from the sugar. Seed crystals, typically sugar from C4 plants, are added to enhance sugar crystallization during boiling. Although comprehensive studies on the health effects of these additives are lacking, their unintended negative impacts on health may affect the coconut sugar market's prospects. Hence, a clean processing method without additives is crucial to preserve the natural quality and properties of coconut sugar.

The conventional open-heat evaporation method often leads to overcooked sugar, resulting in both physical and chemical quality deterioration (Apriyantono et al., 2002). This issue is likely caused by prolonged cooking at high temperatures, which promotes non-enzymatic browning (the Maillard reaction) and results in dark-coloured coconut sugar, typically preferred only as an ingredient in traditional dishes. During the sugar preparation process, the levels of glucose, fructose, and sucrose decrease. After 90 min of boiling, when coconut sugar is formed, the remaining amounts of glucose, fructose, and sucrose are 27.7%, 35.7%, and 83.7% of their initial levels, respectively.

3.2. Modified Conventional Method

In the modified conventional method, coconut sap collected utilizing a Cocosap chiller is concentrated by evaporating moisture at 90°C using a specially designed double-jacketed cooker (Hebbar et al., 2020). To ensure efficient and uniform heat transfer, a thermic fluid fills the space between the cooker's two layers. During continuous heating, the sap undergoes various consistency changes: from an initial pH of 6.7 and 16–18° Brix, it progresses to a syrup stage at 50° Brix, and eventually reaches a thick syrup stage (60–70° Brix). Further heating leads to boiling, and with continuous stirring, the product becomes highly viscous until crystallization occurs. The resulting sugar, with 3 to 4% moisture, forms clods that are rubbed and sieved through a 60-mesh sieve to obtain a sugar product with uniformly sized particles.

Irundu et al., (2022) outlined that the process of making palm sugar involves four key steps: cooking the sap, crystallization, sifting, and drying. In the coconut sugar processing industry, primarily two methods are used: conventional and modern (machine-assisted). Although the stages in both methods are alike, the tools utilized differ. The effectiveness of palm sugar production is assessed based on yield, moisture content, and production time. The conventional method achieved the highest yield percentage of 14.51. The moisture content of sugar obtained from the modern method (2.72%), complies with the ISO standards. Additionally, the modern method saves 130 minutes in production time compared to the conventional method.

3.3. Microwave Technology

Microwave technology provides faster, more uniform heating compared to conventional methods, offering a clean and environmentally friendly process with rapid energy transfer (Leonelli and Mason 2010). While it can be used for concentrating sap syrup, it is unsuitable for sugar crystallization, which requires constant stirring during the final moisture vaporization stage. Tiarasari et al., (2023) reported the optimum drying approach of 80℃ for a drying load of 15 kg while using a specified air oven dryer.

3.4. Rotary Evaporation Technique

Rotary evaporation can be used for sap or *neera* concentration, similar to its applications in chemical, pharmaceutical, and food industries. It offers rapid, gentle, and low-temperature evaporation, minimizing thermal decomposition (Chatterjee et al., 2019). A comparative analysis of *neera* syrup production methods to achieve 65°Brix revealed that rotary evaporation (RE-60) at 60°C and 250 mbar vacuum is the quickest (12.2 min.) and uses the lowest processing temperature (54.8°C). In contrast, microwave evaporation (ME) at 440 W and open heat evaporation (OHE) took 13 min. at 103.2°C and 46.8 min. at 101.6°C, respectively (Asghar et al., 2020b). Additionally, RE-60 syrup is lighter in colour (L* value of 35.17) compared to ME (29.84) and OHE (23.84). The RE-60 syrup is also richer in monosaccharides and requires less energy (0.35 kWh) than OHE (0.83 kWh).

3.5. Drying Method

Coconut sap processed by different methods yields sugar in crystalline or amorphous forms. Crystalline sugar is produced by continuous heating until crystals form, while amorphous sugar is created by placing supersaturated sap in a vacuum or spray drier. Drying coconut sugar powder proves challenging due to its high sucrose content (92%), leading to considerable stickiness resulting from a low glass transition temperature (Tg) (Hebbar et al., 2015). Sucrose possesses a low glass transition temperature (Tg) of 62°C, which is deemed insufficient for the drying process (Bhandari and Howes 1999).

3.6. Vacuum Drying

Vacuum and spray drying are impractical for coconut sugar with a low Tg. To address this, Maltodextrin (MD) is commonly used as a drying agent (Roos and Drusch 2015). MD raises Tg, improves process stability and storage, reduces caking and stickiness, and enhances flowability (Nurhadi and Roos 2017). Among the various drying methods, traditionally produced coconut sugar powder has a higher recovery rate than vacuum-dried and spray-dried powders. Traditional coconut sugar powder is created through the slow crystallization of sucrose due to the low rate of water evaporation (Starzak and Mathlouthi 2010).

Traditionally, coconut sugar is produced by evaporating sap in an open pan over a wood-fired stove until it becomes concentrated. This heating process alters the sugar's unique flavour and colour, impacting its quality during storage. An alternative method involves vacuum evaporation at low temperatures, which helps reduce thermal degradation of food properties. Given the importance of quality in food preparations, it is crucial to control thermal deterioration during processing and storage. Vacuum drying can be employed to produce dried sugar and varying the temperature and drying time can affect the physicochemical properties of the resulting powders. For instance, drying palm sugar at 40°C for 3 h resulted in the highest total phenolic content and antioxidant activity, indicating that vacuum drying can help preserve these beneficial compounds (Aeimsard et al., 2015).

Nurhadi et al., (2018) formulated two different ratios of coconut sugar using sap and MD (7:3 and 6:4 w/w), employing vacuum oven (70°C for 6 h) and spray drying methods (Tinlet 120°C). The spray drying method involving 6:4 ratio of sap and MD yielded the highest monolayer water content and produced a sugar with the brightest and whitest colour. Additionally, the vacuum-dried sugar powder dissolved more rapidly than the spray-dried sugar. The higher concentration of coconut sap to MD (6:4 w/w) in vacuum drying resulted in a powder recovery of 88.70%. Likewise, Nakenan et al., (2013) used a vacuum dryer at 70°C and 80°C to concentrate sap into palm syrup with a Brix level of 70° and found that vacuum drying enhanced the quality of the sugar syrup compared to the open pan method.

Nurhadi et al., (2020) assessed both the conventional and vacuum drying methods for producing coconut sugar. The sugar produced exhibited high hygroscopicity due to its high sucrose content, which makes it prone to caking during storage. To counteract this, an anti-caking agent such as tricalcium phosphate (TCP) was added to maintain stability. The addition of the anticaking agent decreased the monolayer water content and improved the flowability of both types of coconut sugar. The sugar produced by the conventional method predominantly exhibited a crystalline structure, while the vacuum-dried sugar had an amorphous structure

3.7. Spray Drying

Sugars are the main component of coconut sap, with 47.73% sucrose, 11.83% glucose, and 11.53% fructose. However, handling coconut sap during mass sugar production is challenging due to its viscous and sticky nature. As a result, there is increasing demand from consumers and the food industry for dried coconut sap sugar powder. Jayasundera and Kulatunga (2014) developed coconut sap powder using spray drying with a formulation of coconut sap, maltodextrin, and water. The inlet and outlet temperatures were set at 165°C and 65°C, respectively. The resulting spray-dried powder was spherical, with the best recovery (58.6±3.5%) achieved using a 35:15:50 ratio. The initial moisture content and water activity were 1.04 ± 0.003% and 0.43 ± 0.00%, respectively. The total sugar content was 88.2%, comprising 46.6% sucrose, 23.0% glucose, and 18.6% fructose. The powder could be stored for up to 12 months without any changes in moisture or water activity.

Inlet air temperature and maltodextrin concentrations are critical factors affecting the physical properties of coconut sugar powder. A-sun et al., (2016) examined the impact of inlet air temperatures (120, 150, and 180°C) and MD concentrations (10, 20, and 30% w/v) on spray-dried coconut sugar powder. Higher temperatures darkened the powder while reducing moisture, water activity, bulk density, and solubility. Maltodextrin minimized the stickiness, improving spray drying. Higher MD concentrations increased lightness, bulk density, and flowability but decreased solubility and wettability. The optimal condition for coconut sugar powder production was found to be 120°C with 20% MD, yielding good appearance and solubility.

Coconut sugar produced via traditional methods was the least hygroscopic (1.21×10^{-4} g water g^{-1} solid min^{-1}), followed by vacuum-dried (1.48×10^{-4} g water g^{-1} solid min^{-1}) and spray-dried sugar ($1.56\text{-}1.67 \times 10^{-4}$ g water g^{-1} solid min^{-1}). Spray-dried sugar was the brightest and whitest, while vacuum-dried sugar dissolved the fastest (13.33-16.67 seconds). Although increasing maltodextrin in spray drying did not reduce dissolving time (32.00-72.67 seconds), it did enhance the stability of coconut sugar powder, resulting in higher monolayer moisture content, lower hygroscopic rate, and reduced amorphous content.

Traditional coconut sugar powder is made through slow crystallization of sucrose as water evaporates from the sap, leading to co-crystallization and a more crystalline structure (Sekhon 2009). This high crystallinity makes traditional coconut sugar powder more stable. Traditional methods also yielded the highest recovery rate, nearly 100%, while vacuum-dried powder with higher maltodextrin concentrations had a recovery rate of 88.70%, compared to 54.42% for spray drying.

4. Varietal Influence on Sugar Production

Tulalo and Mawardi (2018) evaluated three dwarf coconut varieties - Waingapu Red Dwarf (WRD), Salak Green Dwarf (SGD), and Raja Brown Dwarf (RBD) - at the Indonesian Palma Crops Research Institute to assess their sap and sugar production potential. Sap production per palm per day varied across varieties, with WRD producing the most sap (1.01 l), followed by SGD (0.74 l), and RBD (0.63 l). When processed into crystal sugar, the yield ranked as RBD (16.58%) > SGD (14.38%) > WRD (14.25%). The pH of the crystal sugar ranged from 5.5 (SGD) to 5.6 (WRD and RBD), and the highest sugar content (°Brix) was recorded in RBD (16.63), followed by WRD (16.05), and SGD (14.47). The average tapping duration per inflorescence was 22-23 days across varieties.

Novarianto et al., (2021a; 2021b) examined the sap and sugar production potential in nine dwarf coconut varieties. The sap tapping duration ranged from 31.42 days (SGD) to 43.21 days (WRD) per inflorescence. Sap volume varied from 1.28 l (Jombang Green Dwarf) to 2.6 l (WRD) palm^{-1} day^{-1}, while pH levels ranged from 6.80 (Nias Green and Yellow Dwarfs) to 7.40 (Sagerat Orange Dwarf). The °Brix values spanned from 14.20 (Nias Yellow Dwarf) to 15.49 (Jombang Green Dwarf). Coconut sugar production palm^{-1} day^{-1} ranged from 0.22 kg (RBD) to 0.45 kg (WRD), and the sugar yield varied between 15.57% (RBD) and 17.29% (Jombang Green Dwarf). The dwarf varieties with the highest sap and sugar production potential were WRD, SGD, and Nias Yellow Dwarf, with potential monthly sugar yields of 2.09, 1.64, and 1.56 t ha^{-1}, respectively.

5. Types of Coconut Sap Sugar

5.1. Coconut Sugar Blocks

Coconut sugar blocks are made by heating large quantities of fresh sap, collected in bamboo tubes or other containers, in a large wok at a temperature of around 97°C to 124 ± 2°C for 4 to 4½ h. The sap is stirred occasionally during heating to reach a viscous, thick consistency with a °Brix of 85. After heating, the syrup is cooled for 10 min with continuous stirring before being poured into moulds to solidify into coconut sugar blocks (Iskandar et al., 2014).

Zuliana et al., (2016) identified three main types of coconut sugar products: coconut palm sugar, printed sugar (commonly called brown sugar), and liquid sugar. The colour of coconut sugar varies from yellowish-brown to dark brown, depending on the raw materials and production methods. Coconut sugar has a distinctive taste that sets it apart from other sugars, often used for flavour and to impart a chocolate-like colour. However, printed and liquid coconut sugars have a shorter shelf life of about 2-4 weeks, often lack proper packaging, and are less convenient to serve. In contrast, coconut palm sugar is dried through crystallization, extending its shelf life and making it easier to serve and pack, with the added benefit of easy dissolution in powdered form.

Nusaa and Rimbawan (2022) analyzed the energy levels of different coconut sugar types, ranking them as crystal sugar > block sugar > liquid sugar, with energy values of 377.0, 363.1, and 293.7 kcal 100 g^{-1}, respectively. Crystal coconut sugar also had the highest fat (0.56%), protein (1.29%), and carbohydrate (91.7%) content, followed by block sugar with 0.46% fat, 1.2% protein, and 88.55% carbohydrates. Liquid coconut sugar had the lowest nutritional values in each category. While there were no significant differences in aroma and texture among the three forms, crystal coconut sugar was preferred due to its fine texture, typical brown sugar aroma, dark to light brown colour, and moderately sweet taste. This form of coconut sugar meets the quality standards of palm sugar (SNI 01-3743-1995).

5.2. Granular Coconut Sugar

Crystal coconut sugar is made from coconut sap, which is either processed into powder or crystals or produced by re-melting molded coconut sugar with the addition of water at specific concentrations (Abdullah et al., 2014; Fadhillah et al., 2020). The process of making granular coconut palm sugar is similar to producing coconut sugar blocks, but heating continues until the sap reaches more than 85° Brix (Iskandar et al., 2014). Heating is stopped when the syrup reaches 85-90° Brix, followed by cooling with continuous stirring. As the syrup thickens, stirring speed is increased to form granules, which are then sieved to achieve uniform particle size, resulting in high-quality sugar.

Crystal coconut sugar is known for its lower GI and higher fiber and vitamin B12 levels compared to cane sugar. It is more soluble, has a longer shelf life, and is easier to pack and transport. Its unique taste and aroma make it suitable replacement for cane sugar in products like tamarind syrup, traditional drinks, instant powder drinks, chocolate drinks, coffee mixes, and ice cream. The sugar's quality largely depends on the raw coconut sap, and its powder or fine granules allow for easy dissolution, packaging, and longer storage due to low moisture content. The best organoleptic properties were achieved with sap at pH 7.0 and 1.25% sodium bicarbonate, while the best physico-chemical properties were obtained at pH 7.5 and 0.75% sodium bicarbonate (Zuliana et al., 2016).

Muriel et al., (2019a) reported that freshly collected sap can be transformed directly into granulated sugar through thermal spraying, without preservatives. The crystalline sugar's ash content ranged from 1.45 to 2.85%, titratable acidity from 1.25 to 2.19%, and reducing sugars from 6.75 to 7.89%. Total sugar content represented 81.15 to 87.54% of dry matter, characterized with the sugar darkening as cooking temperatures increased. Cooking 6.25 l of sap produced 1 kg of crystalline sugar (Muriel et al., 2019b). Further studies found that coconut sugars contain significant levels of polyphenols, ranging from 34.64 to 143.12 mg 100 g^{-1}, and granulated coconut sugar was better accepted than white or brown cane sugar (Muriel et al., 2020a; 2020b). Sugars produced at 60-120°C for 45 min were white with a subtle coconut aroma, while sugars at higher temperatures were bland or dark brown with a more pronounced coconut aroma. Coconut sugar's unique properties make it ideal for use in pastries, cakes, ice cream, sweet culinary recipes, and drinks.

Colour plays an important role in consumer acceptance, and granulated palm sugar, being hygroscopic, can darken with increased moisture content. Kurniawan (2020) observed that moisture content in granulated sugar stored at 30°C and 76% RH increased from 3.84% to 8.40% within 10 h, causing the sugar to turn dark brown as the L value (lightness) decreased. Storing granulated coconut sugar with less than 2-4% moisture content helps maintain its quality (Asghar et al., 2021).

Kowsalya et al., (2022) experimented with different energy levels (800, 1000, and 1200 W) for heating inflorescence sap, resulting in final moisture contents of 4.61%, 3.67%, and 2.42%, respectively (initial sap moisture content was 82.01%). Higher energy levels produced sugar granules with lower moisture content, enhancing colour and rehydration properties like wettability, dispersibility, and solubility. The TSS of the sugar also changed more significantly at higher energy levels.

Kencana et al., (2023) used a controlled heating method with a crystallizer machine, involving sap filtration, cooking, crystallization, sieving, and drying, while varying pH and temperature. The produced sugar met Indonesian National Standards (SNI 3743:2021), with 86.93% sucrose content, 1.03% moisture content, and 1.65% ash content. The optimal conditions were found at 110°C and pH 7.0, with the addition of 0.25 g $Ca(OH)_2$.

6. Storage of Coconut Sugar

Quality is crucial in food products, and controlling deterioration during storage is essential. Non-enzymatic browning can negatively impact the nutritional and sensory qualities of stored foods, often limiting shelf life (Apriyantono et al., 2002). Therefore, understanding the processing methods that influence these browning reactions is imperative to preserving the properties of coconut palm sugar syrup during production and storage. This knowledge can be used to optimize thermal processes and minimize quality loss, benefiting both producers and consumers (Naknean et al., 2013).

Gunjal and Galakatu (1986) recommended storing coconut sugar blocks at a relative humidity of 40-45% to maintain quality. Uppal and Sharma (1999) found that the sweetness, colour, and crystallinity of sugar blocks were preserved when stored in airtight containers during the rainy season. Ramya et al., (2010) noted that sugar blocks with higher moisture content and open storage conditions deteriorated quickly, making them unsuitable for consumption. It was also suggested that wrapping sugar blocks in alkathene film, covering them in plastic pouches, and storing them in paper boxes can preserve quality. Drier, harder sugar blocks kept in airtight, hygienic containers maintained even better quality over extended periods.

Trinidad et al., (2015) observed that after six months of storage, the pH of coconut sugar dropped from 5.79 to 5.11, and the °Brix value decreased from 98.9 to 97.6. During this period, water activity increased from 0.44 to 0.54, and titratable acidity (as % lactic acid) rose from 0.3% to 0.5%. The mineral composition of coconut sugar (mg 100 g^{-1}) remained relatively stable, with potassium (K) ranging from 954 to 1075, sodium (Na) from 99 to 112, and iron (Fe) from 0.5 to 0.6. Calcium (Ca) and zinc (Zn) levels stayed constant at 8 and 0.1, respectively. Vitamin C content ranged from 16 to 44 mg 100 g^{-1}.

Coconut sugar powder is typically produced by heating coconut sap until it reaches saturation, resulting in a crystalline structure. Alternatively, spray drying or vacuum drying can be used, producing an amorphous structure (Nurhadi et al., 2018). Due to its high sucrose content, coconut sugar powder from both vacuum drying and conventional methods is highly hygroscopic, making it prone to caking during storage. Adding an anticaking agent like TCP (up to 1% of the total weight) helps stabilize the sugar. TCP effectively reduces hygroscopicity by covering the surface of the crystalline sugar more efficiently than the amorphous version, without affecting the colour or flowability of the sugar (Nurhadi et al., 2020).

7. Coconut Sugar Derived Food Products

Coconut sugar is highly valued for its numerous health benefits, making it an excellent candidate but costly natural sweetener (Hebbar et al., 2022). There is significant demand for coconut sugar in the food and beverage industry, particularly in baked goods like biscuits, and in the production of chocolate, cakes, sweet soy sauce, food coatings, ice cream, and traditional Bengali sweets. Coconut sugar can also be used to create a variety of coconut-based products as detailed below.

7.1. Kalpa Bar

This chocolate, infused with coconut sugar and exclusively made from plant-based ingredients, is entirely dairy-free. The dark chocolate formulation with coconut sugar was optimized to include 45% cocoa liquor, 30% coconut sugar, and 25% cocoa butter. This product is rich in polyphenols, flavonoids, antioxidants, and minerals (Shetty et al., 2023).

7.2. Kalpa Drinking Chocolate

A delicious drinking chocolate prepared from low GI coconut sugar and fine cocoa powder. It releases a rich chocolate aroma when mixed with hot or cold milk.

7.3. Kalparasa-Based Milk Sweets

The concentrated form of Kalparasa is directly used in West Bengal sweets as a substitute for other sweeteners (Ghosh et al., 2016). These sweets provide essential minerals, vitamins, and valuable fibers, and have a low GI, making them a healthy and sweet indulgence.

7.4. Frozen Coconut Delicacy

This product is entirely natural and composed of coconut ingredients such as coconut milk, coconut sugar, tender coconut water, and pulp. The addition of coconut sugar enhances the flavour and taste of the delicacy (Beegum et al., 2022). It is suitable for individuals with lactose intolerance since it contains no dairy ingredients. Replacing refined sugar with coconut sugar lowers the glycemic index and increases the richness of vitamins and minerals.

7.5. Coconut *Peda*

Peda stands out as a highly favored Indian sweet crafted from milk and sugar. Coconut *peda* is prepared using low-fat coconut milk, which is condensed to achieve the khoa stage. This condensed mixture is then combined with coconut sugar, and once it readily separates from the walls of the *khoa* kettle, it is shaped into pedas. The comparison between coconut milk *peda* made with refined sugar and coconut sugar demonstrated a preference for coconut sugar, which also exhibited higher levels of minerals (ash) and protein content (Manikantan et al., 2020).

7.6. Instant *Kheer* Mix

Kheer, a basic and traditional dish, is versatile and enjoyed by people of all communities in any situation. The coconut based instant *kheer* mix optimized at the ratio of 1.5:1.5:1 using coconut milk powder, vermicelli and coconut sugar respectively. The standard amount of water for reconstituting the *kheer* was set at a ratio of 6:1.

7.7. Kalparasa-Based Healthy Coconut Chips

Coconut chips is one of the important coconut-based value added products gaining popularity due to its taste and texture. The dried coconut chips along with osmotic dehydration constitute an effective method for production of superior quality chips. Replacing refined sugar with Kalparasa as an osmotic medium resulted in markedly higher levels of ash, phenolics, and antioxidant activity compared to chips made solely with a sugar solution.

7.8. Kalparasa-Based Nutribar

A high energy nutritional bar can be created by thickening Kalparasa to a concentration of 75–80°Brix (80% concentration), adding protein-rich components like roasted peanuts and green gram, as well as cocoa powder, and then shaping the mixture into bars.

8. Health Benefits of Coconut Sugar

Coconut sugar, though relatively expensive, is often considered one of the best natural sweeteners due to its numerous health benefits (Hebbar et al., 2022). The process of boiling coconut sap to evaporate water results in the formation of brown-coloured sugar, enriched with antioxidant compounds. Despite the high heat treatment involved in its production, coconut sugar retains antioxidant activity, primarily due to the formation of melanoidin. According to Swasti et al., (2024), the antioxidant activity of coconut sap and sugar were measured at 332.51 ppm and 55.37 ppm, respectively, with the high antioxidant level in sap attributed to its ascorbic acid content (Xia et al., 2011).

One of the significant advantages of coconut sugar is its naturally low GI feature, making it beneficial for weight control and regulating glucose and lipid levels in people with diabetes (Njugu, 2013). Coconut sugar's slow energy release helps maintain stable blood sugar levels, avoiding the highs and lows associated with regular sugar consumption. The total sugar content in coconut sugar is around 80% of TSS. Moreover, coconut sap is rich in vitamin C (116.19 ± 0.87 mg l^{-1}) and phenolic compounds (0.34 g GAE l^{-1}), including gallic acid and protocatechuic acid, contributing to its antioxidant activity of 21.85% (Asghar et al., 2020a; Devi et al., 2015).

However, the heat treatment used in coconut sugar production can lead to the formation of compounds, including 5-hydroxyethylfurfural (HMF) and furfural. Swasti et al., (2024) found that while these compounds are present in coconut sugar (HMF at 3.25 ppm and furfural at 462.03 ppm), their levels are below harmful thresholds. Furfural is produced due to glucose present in the sap, and its levels can increase with the higher sucrose content in coconut sap (Rasmussen et al., 2014). Importantly, no furfuryl alcohol was detected in the coconut sugar, reducing concerns about its potential mutagenic effects. In conclusion, while coconut sugar production involves processes that generate some mutagenic compounds, it retains substantial health benefits, such as its antioxidant properties, low GI, and rich nutrient profile, making it a valuable alternative to regular sugar.

Coconut sugar has a sweetening potential similar to saccharose, it is used as an alternative sweetener in the production of confections, beverages, pastries, and other culinary delicacies (Okoma et al., 2020). Trinidad et al., (2010) stated that coconut sugar has a low GI, ranging from 35 to 54 per serving. Low GI diets reduce the risk of developing chronic diseases such as type II diabetes. Coconut sugar is nutritionally superior compared to other sugars. When mixed with wheat flour, sorbitol exhibits the highest starch digestibility, with palm and coconut sugars having nearly the same digestibility (Srikaeo and Thongta 2015).

Renjith et al., (2013) indicated that the methanol extract of inflorescence has excellent antioxidant properties and significant potential for free radical scavenging. This extract was also highly effective in controlling streptozotocin (STZ)-induced hyperglycemia. This effect could be attributed to the presence of dietary fibers, polyphenols, and amino acids such as leucine, arginine, and isoleucine, which possess antidiabetic properties (Doi et al. 2007; Salil et al., 2011). Devi et al., (2015) stated that coconut palm sugar presents a promising alternative to traditional cane sugar owing to its low GI. Additionally, it can serve as a therapeutic agent in the management of type II diabetes, given its αamylase content. The primary sugar component of coconut palm sugar is sucrose, which is comparable to cane sugar. However, coconut sugar differs from cane sugar in that it demonstrates a slower starch digestion rate and lower GI values. These characteristics are attributed to the presence of phytonutrients such as inulin, mineral composition, and its natural, minimally processed form (Srikaeo and Thongta 2015). Multiple compounds found in coconut palm sugar, including polyphenols, inulin, and a high mineral content, have the potential to function as antioxidants. These compounds may help regulate blood sugar levels and enhance lipid balance (Shetty et al., 2023).

9. Browning Intensity of Coconut Sugar

The brown colour is generally used as an indication to assess the quality of coconut sugar. The brown colour of sugar comes during sap heating from non-enzymatic browning reactions known as the Maillard reaction and caramelization (Purnomo 2007). These reactions involve the condensation of an amino group with a carbonyl group (Provost 2019). Coconut sap is rich in reducing sugars and amino acids, which serve as substrates for the Maillard reaction. The primary factors influencing the extent of the Maillard reaction include temperature and time, which are determined by processing conditions. Additionally, pH, water activity, and the type and availability of reactants also play significant roles. The browning intensity, total phenolic contents, and antioxidant activity of coconut sugar increase with a higher initial pH of coconut sap and higher heating temperatures.

The sugar structure present in coconut sap - whether in mono-, oligo-, or polysaccharide form - along with its reactivity with amino groups, plays a crucial role in the colour development of coconut sugar. Studies indicate that the Maillard reaction is significantly responsible for the browning that occurs during the evaporation of coconut sap, contributing to the distinctive flavor and color of coconut sugar. The intensity of the Maillard reaction is primarily influenced by the composition of reducing sugars and amino acids in the sap (Nagi et al., 2018) as well as the heating temperature (Carciochi et al., 2016). In the context of the Maillard reaction, both amino acids and reducing sugars are critical for the development of flavour, colour, and antioxidant properties in food (Wijewickreme et al., 1999). The process is further enhanced by the alkaline pH and elevated temperatures during the heating of coconut sap (Asikin et al., 2014; Ho et al., 2008).

The pH and temperature during the heating of coconut sap significantly influence the intensity of browning, total phenolic content, and antioxidant activity of the resulting coconut sugar. Karseno et al., (2018) found that the highest total phenolic content (0.48%) and browning intensity (0.35) were observed at a pH of 8.0 and a temperature of 115°C. This particular combination also exhibited a notable antioxidant activity, with a DPPH scavenging activity of up to 40%. Thus, there is a strong correlation between browning intensity and antioxidant activity in coconut sugar, highlighting the importance of optimizing processing conditions to enhance the quality and health benefits of the final product.

10. Quality of Coconut Sugar

The quality of coconut sugar is influenced by several critical factors, including the freshness of the palm sap, the cooking process, and the use of preservatives. In the marketplace, the quality of coconut sugar is primarily assessed based on its color and shelf life. High-quality coconut sugar typically exhibits a reddish-brown hue, while lower-quality varieties may appear dark (blackish) or white.

Good-quality coconut sugar is produced from fresh and mature coconut sap that is processed promptly to minimize spoilage and is free from impurities. Conversely, low-quality coconut sugar can spoil rapidly if contaminated or improperly processed. For instance, coconut sap containing more than 8% reducing sugars (as a percentage of total sugars) is deemed unsuitable for crystalline sugar production, leading to a significant yield reduction of up to 50%. However, higher percentages of inverted sugars can be beneficial, particularly in the production of golden syrup (Samarajeewa and Wijeratne 1979).

Coconut sugar sourced from Kulon Progo, Indonesia, is of notably good quality due to its brownish-yellow colour and durability, even when not wrapped or exposed to external air (Lucia et al., 2020). This highlights the importance of specific production conditions and raw material quality in determining the final product's acceptability and shelf life. Kusumawaty et al., (2012) conducted research on the perceived quality of coconut sugar among producers, traders, and buyers in Indonesia. The quality of the sugar produced is influenced by several factors, including the quality of the coconut sap, the use of preservatives (such as sodium metabisulphite and resak wood), technical capabilities, and weather conditions. From the producers' perspective, the colour of their sugar varies significantly, though most prefer pale-coloured coconut sugar, which is typically yellowish-brown (40%) or reddish-brown (38%). The quality of the sugar produced will be influenced by the production conditions (Pratama et al., 2015). A very strong caramel smell and taste is characteristic for coconut blossom sugar (Wrage et al., 2019).

Rakphon and Srijesadaruk (2017) conducted a study on the effects of heating time and temperature on the quality parameters of coconut sap. As heating time increased at both 110°C and 120°C, the moisture content and water activity of the sap decreased, indicating that prolonged heating effectively evaporated water and concentrated the sugars. Conversely, the TSS increased with heating time, demonstrating that the sap became more concentrated in sugars and other dissolved substances. The study also revealed changes in colour parameters during the heating process. The L* value, which represents lightness, decreased, indicating that the sap darkened over time. In contrast, the a* and b* values, which represent redness and yellowness respectively, increased. This trend, along with an increase in the browning index and intimidated browning product values, signifies the development of color compounds typically associated with the Maillard reaction. Additionally, they also noted that most volatile compounds present in the sap increased with extended heating times. Key volatile compounds identified at both heating temperatures included 2,3-dimethyl pyrazine, 2,5-dimethyl pyrazine, 2-methyl pyrazine, and 5-methyl furfural, all of which contribute to the characteristic aroma and flavour of coconut sugar.

Haryanti et al., (2020) corroborated these findings, noting that the colour of coconut sap became darker as heating time increased. They observed a reduction in reducing sugars and free amino acid content during the initial heating phase; specifically, reducing sugars decreased from 3.78% to 1.73% (on a dry basis), and free amino acids dropped from 0.09% to 0.04% (on a dry basis) within the first 10 min. This decrease in amino acids indicated the onset of the carbonyl-amino reaction, which is the initial step in the Maillard reaction responsible for the browning, flavour, and aroma development in heated coconut sugar products. Overall, these studies highlight the significant impact of heating parameters on the quality characteristics of coconut sugar, particularly regarding moisture content, colour development, and volatile compound formation, all of which contribute to its sensory attributes.

According to the Philippines National Standard (PNS/BAFPS 76:2010 ICS 67.180), the coconut sap sugar should conform to the requirements specified in Table 2.

Table 2. Coconut sap sugar – Grading and classification (page 2)

Class	Specification
Premium class (Superior quality)	Colour: Light yellow to cream; Water Activity (aw) < 0.5
Class I (Good quality	Colour: Light brown to brown; Water Activity (aw) 0.5-0.7
Class II	This class includes coconut sap sugar which does not qualify for inclusion in the higher classes, but satisfy the minimum requirements specified under coconut sugar quality standards

As per the PNS/BAFS 76:2016 ICS (Coconut Sap Sugar – Specifications) (Bureau of Agriculture and Fisheries Standards), the details for physical, chemical and microbiological characteristics are as follows (Table 3).

The Indonesian government sets standards for the quality of coconut sugar, which encompass criteria such as physical appearance, aroma, taste, and chemical composition including water content and sucrose levels.

The Indonesian National Standard SNI 3743: 2021 standard (Components of palm sugar) is as follows (Table 4).

Table 3. Physical, chemical and microbiological characteristics of coconut sap sugar)

Physical characteristics	
Parameter	Quality characteristics
Colour	Light yellow/cream to dark brown
Odour	Sweet scent; pleasant nutty aroma
Taste	Sweet
Others	Free from filth and extraneous matters
Chemical characteristics	
Parameter	Values (%)
Water activity (aw)	$\leq$0.6
Moisture content (%)	$\leq$ 3.5
Ash	$\leq$ 2.4
Glucose	2.0-3.0
Fructose	1.0-4.0
Sucrose	78.0-89.0
Microbiological characteristics	
Salmonella (/25g)	Negative
E. coli (MNP/100g) or (cfu g^{-1})	Negative
Coliform Count (cfu g^{-1})	< 10 (cfu g^{-1}) or < 3 MPN/g
Total Plate Count (cfu g^{-1})	<500
Mold and Yeast (cfu g^{-1})	<10

Table 4. SNI 3743: 2021 Components of palm sugar

Component	Content
Reduced Sugar (%)	Max 3,0
Sucrose (%)	80- 93
Moisture (%)	Max 3.0
Ash (%)	Max 2.5
Water insoluble part (%)	Max 1
Colour Substance	Allowable
Lead (Pb)	Negative
Starch	Negative
Shape	Powder

The International Coconut Community (ICC) Standard for coconut sugar (https://apccsec.org/public/products-detail/coconut-sugar) is given in Table 5.

Table 5. ICC Standard for coconut sugar

Parameters	Value
Colour	Light yellow/cream to dark brown
Odour	Sweet, pleasant nutty aroma
Taste	Sweet
Others	Free from filth and extraneous matters
Moisture content (%)	Cube/block: Max 6; Granular: Max. 3
Glucose (%)	2.-3.0
Fructose (%)	1.0-4.0
Sucrose (%)	78.0-89.0
Ash (%)	≤2.4
Microbiological contaminant	
Total Plate Count (cfu g^{-1})	Cube/block: 10,000; Granular ≤10
Coliform (cfu g^{-1})	Cube/block: 100; Granular <10
E. coli	Negative
Salmonella	Negative
Yeast/molds (cfu g^{-1})	<10
Heavy Metals (max.mg/kg)	
Zinc (Zn)	40.0
Lead (Pb)	2.0
Copper (Cu)	10.0
Mercury (Hg)	0.03
Tin (Sn)	40.0
Arsenic (As)	1.0

Generally, coconut sugar producers believe that high-quality coconut sugar should have a tough and sandy texture. The tough texture may impact shelf-life as it doesn't dissolve rapidly, while the sandy texture leads to a brittle consistency and facilitates easy cutting. Regarding aroma and flavour, producers view the characteristic and natural fragrance and sweetness of coconut sugar as indicators of its quality.

Hygiene plays a crucial role in coconut sugar production, with cleanliness of the sap-tapping container being paramount for ensuring quality sugar. While most producers emphasize the need for coconut sugar to be free from contamination, they often overlook hygiene during the production process. Many fail to regularly clean the sap container or filter the fresh sap, resulting in contaminants being frequently present in the produced coconut sugar.

11. Applications in the Food Industry and Sustainability Concerns

The GI values of coconut sap and coconut sugars are notably low, with values of 35 ± 4 and 42 ± 4, respectively (Trinidad et al., 2010). These low GI values make coconut sap and sugars particularly suitable for individuals suffering from diabetes, as they provide a more stable energy release compared to high-GI sweeteners. This characteristic positions them as a preferable alternative to synthetic sugars, given their natural origins and traditional production methods. In contrast, a mixture of wheat flour and coconut or palm sugar has a higher GI, reported at around 61 (classified in the medium category), according to an in vitro evaluation conducted by Srikaeo and Thongta (2015). This indicates that while coconut sugar retains its low GI profile, the addition of other ingredients like wheat flour can elevate the overall glycemic response of the mixture.

Recent research by Sawitzki and da Silva (2023) supports the notion that consuming foods with low GI and glycemic load is associated with a reduced risk of chronic diseases. This reinforces the potential health benefits of incorporating coconut sugars and sap into diets, particularly for individuals looking to manage blood sugar levels and enhance overall health. The combination of low GI values and the associated health benefits underscores the value of coconut sugar as a natural sweetener option.

Coconut sugar is produced by concentrating the inflorescence sap, being natural origin and minimally processed, it can be easily incorporated into many vegan diets as a healthier alternative. The production methods of coconut sugar vary significantly based on local, traditional, and indigenous practices (Wrage et al., 2019). These factors contribute to the wide range of appearances, tastes, and flavors found in the various types of coconut sugar available on the market. The traditional method of collecting coconut sap involves placing an open earthen pot or bamboo sac at the top of the palm to collect the sap for 8-12 h. Lime is then applied to the inner surface of the pot to prevent fermentation. This method results in sap that is oyster white in color and has a strong odor, often contaminated with insects, ants, pollen, and dust particles. However, the coco-sap chiller developed by the ICAR-Central Plantation Crops Research Institute (CPCRI) in India has improved the quality of unfermented coconut sap. It has reduced processing time, eliminated contaminants, and enabled better product diversification and market opportunities. Because of its high sucrose content, coconut sugar tends to clump together during storage. To prevent this, an anticaking agent like TCP is often added. TCP coats the surface of the coconut sugar powder, reducing its tendency to absorb moisture and form clumps. This process improves the flowability of the sugar, making it easier to use in various food applications (Swain et al., 2016). Using coconut sugar syrup, which contains high levels of antioxidants, shows promise as a food production ingredient. Previous studies have highlighted that coconut sugar with higher levels of vitamins, minerals, and antioxidants can serve as a natural sugar alternative with enhanced chemical properties (Hebbar et al., 2015).

The use of low GI sugar, like coconut sugar, in the production of plain chocolate shown to be more nutritious, containing higher levels of anti-carcinogenic compounds, antioxidants, and minerals compared to commercial chocolates made with sugar cane or sugar palm. Additionally, incorporating coconut sugar in food processing at high temperatures can enhance antioxidant activity. Significant compounds such as pyroglutamic acid and hydroxymethylfurfural (HMF) are formed when heated (Saputro et al., 2019).

Coconut sugar (Figure 2) is favoured for its ecological benefits and nutritional properties. It is widely used in the food and beverage industries for products such as plain and drinking chocolate, cakes, cookies, and brownies. Additionally, it serves as a sweetener for juices, teas, and other beverages, and can be used as a seasoning agent. The inclusion of coconut sugar in various food applications is well justified as a healthy option due to its rich content of essential nutrients such as vitamins E and C, minerals like zinc, iron, potassium, and phosphorus, and phytonutrients including anthocyanidins, flavonoids, polyphenols, and antioxidants. Moreover, coconut sugar contains a notable amount of inulin (4.7 g $100g^{-1}$), which is essential for the production of short-chain fatty acids like acetate, butyrate, and propionate (Hebbar et al., 2015; Muriel et al., 2021).

Utilizing palm sugar as a sweetener in soybean sauce has a significant impact on the flavour due to the presence of over 70 volatile compounds (Puspitasari-Nienaber and Sumaryanto 1996). Additionally, using palm sugar as a natural sweetener also affects the color, texture, and flavour of cookies. Another study reveals that during baking, bread sweetened with palm sap sugar has a lower GI than bread sweetened with cane sugar, Furthermore, mixture of palm sugars and corn starch results in a slower digestion rate (Srikaeo and Thongta 2015). Coconut sugar holds a prominent place in traditional Indonesian cuisine, appearing as a key ingredient in various dishes and beverages. Notably, it contributes significantly to the distinct taste and flavour of Indonesian sweet soy sauce (kecap manis) and the preparation of Indonesian intermediate moisture meat (dendeng) (Purnomo and Suryoseputro 2001). Coconut sugar plays a crucial role in maintaining the stability of coconut milk. The emulsion stability of coconut milk samples increased with higher levels of coconut sugar content (Jirapeangtong et al., 2008).

Figure 2. Coconut palm sugar (Photo credit: Deepti Nair S).

In terms of sustainability and traceability, organic certification serves as a quality standard that enhances the credibility of coconut sugar in the European market. Export traders can promote the fact that coconut sugar is traditionally produced by small farmers, that palm trees grow organically among other crops, and that the sugar has very low fructose content and lower GI values compared to traditional refined beet or cane sugar. To ensure coconut-based agroecosystems are resilient to climate change, a holistic approach that prioritizes the overall productivity and sustainability of the entire system, rather than just the palms, is necessary. As the demand for natural products continues to grow, the use of alternative sweeteners like coconut syrup and sugar is expected to increase (Saraiva et al., 2023).

12. Detection of Adulteration in Coconut Sugar

The quality of coconut sap is significantly influenced by various factors, including weather conditions, the tapping season, and the duration of tapping (Naknean et al., 2010). During the rainy season, sugar content in coconut sap can drop sharply, complicating the granulation process of coconut sugar (Haryanti et al., 2018). To counteract this issue, sugar manufacturers often add a small amount of cane sugar to enhance the yield of granulated coconut sugar (GCS) and expedite the crystallization process. This practice is similar to that followed in maple syrup production, where the risk of adulteration by adding cane sugar (stretching) is also a concern (Tremblay and Paquin 2007). Typically, less than 5% (w/v) of cane sugar is added to coconut sugar to "seed" and accelerate the crystallization process. While the addition of cane sugar increases GCS production by promoting the formation of crystal nuclei, it can alter the chemical and sensory properties of the final product, making GCS with cane sugar distinct from pure GCS (Wiboonsirikul 2016). As a result, coconut sugar produced with cane sugar additions may not meet the desired quality for export.

In contrast, organic GCS, made exclusively from coconut sap sucrose without any additives, is essential for achieving high export quality. However, studies identifying specific biomarkers for distinguishing cane sugar from coconut sugar are still limited (Jahren et al., 2006; Psomiadis et al., 2018).

Recent advancements in analytical techniques have been proposed to detect the adulteration of coconut sugar. Notably, Nuclear Magnetic Resonance (NMR) and Isotope Ratio Mass Spectrometry (IRMS) are significant methods highlighted in research by Bachmann et al., (2022) and

Rogers et al., (2021). NMR is a non-destructive technique that provides rapid results and requires minimal sample preparation. It can be particularly effective in analyzing the metabolic profiles of coconut sugar (Gowda and Raftery 2015). A major drawback of NMR is the necessity for a larger sample size to obtain a sufficient signal (Meenu et al., 2019). IRMS (Isotope Ratio Mass Spectrometry) requires a smaller sample size than NMR and achieves precision of 0.2‰ on the δ-scale (Gilbert et al., 2012). Despite its efficacy in detecting cane and corn sugar adulteration in coconut sugar and syrup, IRMS cannot identify beet sugar additions due to both coconut and beet being C3 plants, which results in similar δ13C values (Ghidotti et al., 2021; Phaeon et al., 2021). Despite these limitations, both NMR and IRMS have substantial potential for addressing food fraud and adulteration issues in coconut sugar production. Energy-dispersive X-ray fluorescence (ED-XRF) is another promising technique for detecting adulteration in coconut sugar. As demonstrated by Zdiniakova and Calle (2020), ED-XRF can effectively identify adulterations with cane and beet sugars. Its primary benefits include minimal sample preparation and the ability to use portable devices, although it does have relatively high quantification limits. Combining the strengths of different analytical approaches, such as NMR, IRMS, and ED-XRF, could greatly enhance the ability to detect adulteration in coconut sugar, ensuring product integrity and quality for both producers and consumers. This multi-faceted approach will be vital in addressing the growing concerns about food fraud and maintaining consumer trust in natural sweeteners like coconut sugar.

The high price of coconut sugar, attributed to its superior quality and economic value, leads consumers to pay 20 to 65 times more for it than for conventional refined sugar. However, it is common for manufacturers to substitute sugar-rich products with cheaper alternatives without proper labeling (Russell et al., 2021). Examples of such substitutes include starch-derived sugar syrups, high-fructose corn syrup, glucose syrup, and sucrose syrups from sugar beet or cane sugar. These substitutions alter the proportions of glucose, fructose, and sucrose, as well as the $_{13}C/_{12}C$ isotope ratio. While $_{13}C/_{12}C$ isotope analysis can detect the presence of cane sugar (C4 plant) and high-fructose corn syrup (C4), identifying beet sugar (C3) is challenging, posing a significant issue (Tosun 2013).

Wrage et al., (2019) conducted a study on the chemical composition and sensory attributes of 107 samples of coconut sugar, which included 79 samples from Indonesia, 6 from the Philippines, and 22 from unknown origins. The analysis indicated that coconut sugar might also contain gluten, starch, sugars from C4 plants, as well as coconut or palm oil. Their findings revealed that

17% of the samples had a notable presence of C4 sugar (≥ 10 g 100 g^{-1}), while 8% contained gluten at levels of ≥ 20 mg kg^{-1}. These additives could serve technological purposes and enhance the overall yield. Coconut sugar, with its high sucrose content, tends to clump during storage, necessitating the use of an anti-caking agent like TCP. TCP coats the sugar's surface, reducing hygroscopicity and enhancing flowability (Swain et al., 2016). Due to the similar production processes of palm sugar and coconut sugar, distinguishing between the two can be difficult. Coconut sugar retains its quality and nutritional benefits when processed from clean, unfermented sap. However, using low-quality *neera* requires additional chemicals and additives like gluten, starch, and sugars from C4 plants to facilitate crystallization. To prevent overboiling, coconut quarters and oil are added, while starch helps absorb moisture to prevent agglutination (Levang 1988). Unfortunately, these additives may have adverse health effects, potentially impacting the coconut sugar market.

Zdiniakova and Calle (2020) utilized a multivariate analysis method called Soft Independent Modelling of Class Analogies (SIMCA) to examine the elemental composition of eleven coconut sugar samples, ten cane sugar samples, and one beet root sugar sample sourced from supermarkets in Belgium, Spain, Poland, and Italy. They employed energy dispersive X-ray fluorescence (ED-XRF) for this analysis, finding it effective for authenticating coconut sugar due to its sensitivity in detecting essential elements such as P, Cl, K, Ca, Fe, Cu, Br, Rb, and Sr. The study revealed that the average mass fractions of these elements were generally higher in coconut sugar compared to cane and beet root sugars. Consequently, if the mass fractions of these elements in a coconut sugar sample fall below the quantification limits (LoQ) or are outside the expected range, the sample should be considered questionable and may require additional, more specific testing.

Rogers et al., (2021) employed carbon isotope testing to assess 109 coconut sugar samples from Indonesia over two years to determine if they were adulterated with more than 5% C4 cane sugar. The results showed that genuine and unadulterated coconut sugar had an average δ13C value of −25.6‰ ± 0.4‰. Seasonal climatic variations could cause standard deviations in coconut sugar δ13C values of up to ± 0.4‰. It is common for producers to add up to 5% w/v of cane sugar for seeding, which could raise the final δ13C values of coconut sugar by as much as 0.7‰. δ13C values greater than −24.8‰ suggest the presence of C4 sugars, such as cane or corn sugar/syrup, while values lower than −26.4‰ are likely indicative of additives like mangosteen sap, which is used to prevent fermentation in sugar syrup. Overall, the authors recommend a maximum δ13C value of −24.1‰ for authentic coconut sugars. While this carbon isotope ratio analysis method offers high accuracy, it is also quite costly for assessing the purity of coconut sugar.

Organically sourced coconut sugar, known for its distinct caramel aroma, is particularly vulnerable to adulteration. Bachmann et al., (2022) highlight that the high similarity among various types of brown sugar, along with their comparable aroma profiles and significant price differences make coconut sap sugar a potential target for food fraud. Their study examined the metabolic profiles of unrefined cane sugar, browned beet sugar, and coconut blossom sugar using ^{1}H NMR spectroscopy, identifying three marker metabolites to differentiate the three types of sugar. Notably, they did not detect hydroxymethyl furfural (HMF) in any of the samples. However, pyroglutamic acid was recognized as a unique marker for coconut sugar, while the presence of metabolites from other sugars could indicate adulteration. Additionally, coconut sugars were found to have significantly higher levels of acetic, formic, lactic, and succinic acids compared to both cane and beet sugars. Trans-aconitic acid and betaine were identified as markers for cane and beet sugars, respectively. Even in the absence of clear markers, it is possible to detect adulteration of coconut sugar at levels between 10% and 30%. Based on the diverse metabolic profiles, the study demonstrated that a multivariate regression model could effectively estimate the extent of adulteration in coconut blossom sugar. To further enhance the identification of coconut sugar's unique metabolic profiles, Saraiva et al., (2023) recommended combining liquid chromatography-mass spectrometry (LC-MS) with NMR techniques.

Sulistyo et al., (2024) developed a low-cost portable LED-based spectrometer that effectively detects cane sugar in GCS using machine learning (ML) with high accuracy. The first neural network they constructed achieved a classification accuracy of 99.4% in distinguishing between unadulterated and adulterated granulated coconut sugar. The second neural network was able to classify the concentration of cane sugar added to coconut sap (up to 5% w/v) with a classification accuracy of 94.1%.

Ensuring food safety is crucial alongside concerns about food adulteration. Coconut sugar offers valuable nutritional and functional benefits to consumers if produced from high-quality inflorescence sap and processed according to good manufacturing practices (GMP). However, using inferior-quality sap and processing under unhygienic conditions can lead to contamination by insects and microorganisms. Therefore, it is vital to conduct thorough studies to assess potential contaminants. Research on contaminants from food processing (e.g., acrylamide), agrochemicals, heavy metals (e.g., mercury, lead, arsenic, cadmium), microbial toxins (e.g., mycotoxins), and cleaning agents or disinfectants has increased for other natural sweeteners and sugar products but remains largely absent for coconut sugar and syrup (Matondang et al., 2017).

Conclusion

Coconut sugar is often considered a healthier sweetener than most commercially available sugars. The considerable amounts of potassium, vitamin C, inulin, and its fermentation product (propionate) are added benefits of coconut sap sugar, making it a promising functional food or ingredient. In India, small-scale industries in Tamil Nadu, Andhra Pradesh, Kerala, the Lakshadweep Islands, and Karnataka produce and sell coconut sugar. Despite a significant demand for coconut sugar, production and supply remain limited. Recent advancements in the collection of coconut sap and its processing into natural value-added products, along with their associated health benefits, have led to a rapid increase in both domestic and international markets for Kalparasa and coconut sugar.

The farm production of coconut sugar has the potential to be a profitable venture in various coconut-growing countries, provided there is a sufficient supply of raw materials and effective marketing strategies for the product. Conducting more thorough comparative analyses and adopting improved alternative technologies, both technical and financial, to enhance the competitiveness of coconut sugar production are warranted.

References

Abdullah, W.G., Rianse, U., Iswandi, R.M., Taridala, S.A.A., Rianse, I.S., Zulfikar, Z.L., Baka, L.R., Abdi, A.L. (2014) Potency of natural sweetener: brown sugar. *Adv Environ Biol* 12(1):374-386.

Aeimsard, R., Thumthanaruk, B., Jumnongpon, R., Lekhavat, S. (2015). Effect of drying on total phenolic compounds, antioxidant activities and physical properties of palm sugar. *J Food Sci Agric Technol* 1:126-30.

Apriyantono, A., Aristyani, A., Lidya, Y., Budiyanto, S., Soekarto, S.T. (2002). Rate of browning reaction during preparation of coconut and palm sugar. In: International Congress Series. Elsevier, pp 275–278.

Asghar, M.T., Yusof, Y.A., Mokhtar, M.N., Ya'acob, M.E., Ghazali, M.H., Chang, L.S., Manaf, Y.N. (2020a). Coconut (*Cocos nucifera* L.) sap as a potential source of sugar: Antioxidant and nutritional properties. *Food Sci Nutn* 8 (4): 1777–1787.

Asghar, M.T., Yusof, Y.A., Mokhtar, M.N., Yaacob, M.E., Ghazali, H.M., Varith, J., Chang, L.S., Manaf, Y.N. (2020b). Processing of coconut sap into sugar syrup using rotary evaporation, microwave, and open-heat evaporation techniques. *J Sci Food Agric* 100 (10):4012–4019.

Asghar, M.T., Yusof, Y.A., Mokhtar, M.N., Yaacob, M.E., Ghazali, H.M., Chang, L.S. (2021) A review of nutritional facts, production, availability and future aspects of coconut palm sugar. *J Nutr Food Sci.* 11(3):793, 8 pages.

Asikin, Y., Kamiya, A., Mizu, M., Takara, K., Tamaki, H., Wada, K. (2014). Changes in the physicochemical characteristics, including flavour components and Maillard reaction products, of non-centrifugal cane brown sugar during storage. *Food Chem* 149: 170–177.

A-sun, K., Thumthanaruk, B., Lekhavat, S., Jumnongpon, R. (2016) Effect of spray drying conditions on physical characteristics of coconut sugar powder. *Int Food Res J* 23(3): 1315-1319.

Bachmann, R., Horns, A.L., Paasch, N., Schrieck, R., Weidner, M., Fransson, I., Schrör, J.P. (2022). Minor metabolites as chemical marker for the differentiation of cane, beet and coconut blossom sugar. From profiling towards identification of adulterations. *Food Control* 135, 108832. https://doi.org/10.1016/j.foodcont.2022.108832.

Beegum, P.P.S., Nair, J.P., Manikantan, M.R., Pandiselvam, R., Shil, S., Neenu, S., Hebbar, K.B. (2022). Effect of coconut milk, tender coconut and coconut sugar on the physico-chemical and sensory attributes in ice cream. *J Food Sci Technol* https://doi.org/10.1007/s13197-021-05279-y, 13 pages.

Bhandari, B.R., Howes, T. (1999). Implication of glass transition for the drying and stability of dried foods. *J Food Eng* 40 (1–2):71–79.

Carciochi, R.A., Dimitrov, K., D.´Alessandro, L.G. (2016). Effect of malting conditions on phenolic content, Maillard reaction products formation, and antioxidant activity of quinoa seeds. *J Food Sci Technol* 53(11):3978-3985. https://doi.org/10.1007/s13197-016-2393-7.

Chatterjee, S., Sugilal, G., Prabhu, S.V. (2019). Heat transfer in a partially filled rotary evaporator. *Int J Thermal Sci* 142:407–421.

Devi, N.S., Hariprasad, T., Ramesh, K., Merugu, R. (2015). Antioxidant properties of coconut sap and its sugars. *Int J Pharm Tech Research* 8 (1):160–162.

Doi, M., Yamaoka, I., Nakayama, M., Sugahara, K., Yoshizawa, F. (2007). Hypoglycemic effect of isoleucine involves increased muscle glucose uptake and whole body glucose oxidation and decreased hepatic gluconeogenesis. *Am J Physiol- Endocrinol Metabolism* 292 (6): E1683–E1693.

Fadhillah, N., Mela, E., Mustaufik (2020) Crystal coconut sugar and its potential use in beverage products. *Agritech* 12(1):20-28.

Ghidotti, M., Papoci, S., Dumitrascu, C., Zdiniakova, T., Fiamegos, Y., Gutiñas, M.B.C. (2021). ED-XRF as screening tool to help customs laboratories in their fight against fraud. State-of-the-art. *Talanta Open* 3, 100040, https://doi.org/10.1016/j.talo.2021.100040.

Ghosh, D.K., Bandyopadhyay, A., Baidyanath, D., Kaushik, G., Hebbar, K.B., Chowdappa, P. (2016). Method of the preparation of fresh coconut inflorescence sap (Kalparasa) based sweets. Patent Application No. 201631013187.

Gilbert, A., Hattori, R., Silvestre, V., Wasano, N., Akoka, S., Hirano, S., Yamada, K., Yoshida, N., Remaud, G.S. (2012). Comparison of IRMS and NMR spectrometry for the determination of intramolecular 13c isotope composition: Application to ethanol. *Talanta* 99:1035–1039.

Gowda, N.G.A., Raftery, D. (2015) Can NMR solve some significant challenges in metabolomics? *J Magn Reson* 260: 144–160, doi: 10.1016/j.jmr.2015.07.014.

Gunjal, B.B., Galakatu, P.D. (1986). Water vapour adsorption by jaggery. *Indian Sugar* 35:1-4.

Haloho, W.F., Susanto, W.H. (2015) Pengaruh penambahan larutan susu kapur dan stpp (sodium tripolyphospat) terhadap kualitas gula kelapa (*Cocos nucifera* L) (Effect the addition of lime and sodium tripolyphospate (stpp) to quality palm sugar). *J Pangan dan Agroindustri* 3(3): 1160–1170.

Hanifah A.N., Rahayoe S., Saputro A.D., Kusuma R.A. (2022) Kinetics of the coconut sap physical properties during palm sugar processing using pan evaporator and vertical type double jacket stirred crystallizer. *Adv Biol Sci Res* 19:166-173, doi: 10.2991/absr.k.220305.025 .

Haryanti P., Sulistyo S.B. (2022) The effects of adding lysine to sap on chemical characteristics and antioxidant activity of granulated coconut sugar. *Food Res* 6 (3) : 306 – 314, https://doi.org/10.26656/fr.2017.6(3).420.

Haryanti, P., Supriyadi, Marseno, D.W., Santoso, U. (2018). Effects of different weather conditions and addition of mangosteen peel powder on chemical properties and antioxidant activity of coconut sap. *Agritech* 38(3):295–303.

Haryanti, P., Supriyadi, Marseno, D.W., Santoso, U. (2020) The changes of chemical composition and antioxidant activity of coconut sap during heating process. *Rasayan J Chem* 13(3):2010-2019.

Hebbar, K.B., Arivalagan, M., Manikantan, M.R., Mathew, A.C., Thamban, C., Thomas, G.V., Chowdappa, P. (2015). *Coconut inflorescence sap and its value addition as sugar–collection techniques, yield, properties and market perspective.* 109(8):1411–1417.doi: 10.18520/v109/i8/1411-1417.

Hebbar, K.B., Arivalagan, M., Pavithra, K.C., Roy, T.K., Gopal, M., Shivashankara, K.S., Chowdappa, P. (2020). Nutritional profiling of coconut (*Cocos nucifera* L.) inflorescence sap collected using novel coco-sap chiller method and its value added products. *J Food Measurement and Characterization* 14: 2703–2712.

Hebbar, K.B., Ramesh, S.V., Ghosh, D.K., Beegum, P.P.S., Pandiselvam, R., Manikantan, M.R., Mathew, A.C. (2022). Coconut sugar-a potential storehouse of nutritive metabolites, novel bio-products and prospects. *Sugar Tech* 24 (3): 841–856.

Henrietta, H.M., Kalaiyarasi, K., Raj, A.S. (2022). Coconut tree (*Cocos nucifera*) products: A review of global cultivation and its benefits. *J Sustain Environ Manage* 1 (2): 257–264.

Ho, C., Aida, W.W., Maskat, M.Y., Osman, H. (2007). Changes in volatile compounds of palm sap (Arenga pinnata) during the heating process for production of palm sugar. *Food Chem* 102(4):1156-62.

Ho, C., Aida, W.W., Maskat, M.Y., Osman, H. (2008). Effect of thermal processing of palm sap on the physico-chemical composition of traditional palm sugar. *Paki J of Bio Sci*11(7):989- 95.

Indrayanto, A., Suliyanto , Restianto, Y.E., Dinanti, A., Naufalin, L.R., Krisnawati A., Nurchasanah , Iskandar D., Kurniawan R.E.K. (2022) Organic coconut sugar factory: Feasibility study Indonesia case,. *Int J Res* 10(11):140–152, doi: 10.29121/ granthaalayah.v10.i11. 2022.4887.

Irundu, D., Khoiriyah, M., Ritabulan, Ramli M.A., Zulkahfi (2022) Efektivitas pembuatan gula semut menggunakan metode konvensional dan modern (Effectiveness of making palm sugar using conventional and modern methods). *Jurnal Penelitian Kehutanan Bonita* 4(1):30-37.

Iskandar, A., Yuliasih, I., Haryanto, D.B. (2014). Effect of temperature and time on dry granulation process of arenga palm sugar. Impact. IMPACT: *Int J Res Eng Technol*2(9):33- 44.

Jahren, A.H., Saudek, C., Yeung, E.H., Kao, W.H.L., Kraft, R.A., Caballero, B. (2006). An isotopic method for quantifying sweeteners derived from corn and sugar cane. *Am J Clin Nutrn* 84:1380–1384, https://doi.org/10.1093/ajcn/84.6.1380.

Jayasundera, J., Kulatunga, A. (2014). Spray-drying of coconut treacle into an amorphous powder. *Emirates J Food Agric* 26(8):672-678, doi: 10.9755/ejfa.v26i8.17581.

Jirapeangtong, K., Siriwatanayothin, S., Chiewchan, N. (2008). Effects of coconut sugar and stabilizing agents on stability and apparent viscosity of high-fat coconut milk. *J Food Eng* 87 (3): 422–427.

Jose, S.P., Asha, S., Krishnakumar, I.M., Ratheesh, M., Santhosh, S., Sandya, S., Kumar, G., Pramod, C. (2017). Nephro-protective effect of a novel formulation of unopened coconut inflorescence sap powder on gentamicin induced renal damage by modulating oxidative stress and inflammatory markers. *Biomed Pharmacotherapy* 85: 128–135.

Kabir, M.A., Lorjaroenphon, Y. (2014). Identification of aroma compounds in coconut sugar. In: *Proceedings of the 52nd Kasetsart University Annual Conference; Bangkok, Thailand,* 4–7 February 2014; pp. 239–246.

Karseno, E., Yanto, T., Setyowati, R., Haryanti, P. (2018). Effect of pH and temperature on browning intensity of coconut sugar and its antioxidant activity. *Food Res* 2 (1): 32–38.

Kencana, A.B., Dewi, E., Taufik, M. (2023) The impact of acidity (ph) and temperature on the crystallization process of coconut palm sugar. *Equilibrium J Chem Eng* 7(2):199–205, https://doi.org/10.20961/equilibrium.v7i2.79964.

Kowsalya, P., Kamalapreetha, B., Anbarasan, R., Mahendran, R. (2022) Processing of coconut neera to produce sugar granules and study on its properties. *Sugar Tech* 25:603–609. https://doi.org/10.1007/s12355-022-01208-1.

Kroger, M., Meister, K., Kava, R. (2006). Low-calorie sweeteners and other sugar substitutes: a review of the safety issues. *Food Sci Food Safety* 5(2):35-47, https://doi.org/10.1111/j.1541-4337.2006.tb00081.x.

Kurniawan, H. (2020) Pengaruh kadar air terhadap nilai warna cie pada gula semut (Effect of moisture content on CIE color values in granulated palm sugar). *Jurnal Teknik Pertanian Lampung* 9(3):213-221, http://dx.doi.org/10.23960/jtep-l.v9i3.213-221.

Kurniawati, E., Karimah, R.N., Suryana, A.L., Destarianto, P., Oktafa, H. (2021) The effect of sodium metabisulfite concentration and drying time on the quality of coconut sugar. OP Conf. *Series: Earth and Environmental Science* 672, 012069 , doi:10.1088/1755-1315/672/1/012069.

Kurniawati, E., Nurul, K.R., Lironika, S.A., Destarianto, P. (2019) Implementation of good manufacturing practice (GMP) on coconut palm sugar processing at craftsmen business group in Wonosobo-Banyuwangi village as a Helicos center. *Proc. The Second International Conference on Food and Agriculture Bali Nusa Dua Convention center*, 2-3 Nov.2019, pp 127-134.

Kusumawaty, Y., Maharani, E., Edwina, S. (2012). Perceived quality of coconut sugar by producers, traders and downstream industries in Indragiri Hilir District, Riau Province, Indonesia. *J Agribusiness Marketing* 5 (1): 13.

Leonelli, C., Mason, T.J. (2010). Microwave and ultrasonic processing: Now a realistic option for industry. *Chem Eng Processing: Process Intensification* 49 (9): 885–900.

Luchini, P.D., Bettani, S.R., Bernardi, V.M.R., Borges, M. (2017). Mineral and metal levels in brown sugar from organic and conventional production systems. *J Agric Sci* 9 (10): 226–233.

Lucia, K.C., Waluyati, L.R., Suryantini, A. (2020) Supply chain analysis of coconut palm sugar in Kokap subdistrict, Kulon Progo regency. *J Agribusiness Management and Dev*1(1):49-57.

Mammen, R., John, F., Krishnakumar, I.M., Maliakel, B., Mohanan, R., Fleenor, B. (2020). Safety and influence of a novel powder form of coconut inflorescence sap on glycemic index and lipid profile. *Bioactive Carbohydrates Dietary Fibre* 23, 100217, https://doi.org/10.1016/j.bcdf.2020.100217.

Manikantan, M.R., Beegum, P.P.S., Pandiselvam, R., Hebbar, K.B. (2020). Coconut milk Peda. Kalpa CPCRI Newsletter 39 (1): 4.

Matondang, K.A., Suman, A., Sasongko, Mahardika, A.S.P(2017) Local economic development-based on farming manufacture of coconut sugar in West Sei Kepayang district. *Int J Social and Local Econ Gov* 3(1):49-55.

Meenu, M., Cai, Q., Xu, B. A. (2019) Critical review on analytical techniques to detect adulteration of extra virgin olive oil. *Trends Food Sc. Technol* 91:391–408.

Muriel, J.O.D, Louis, K.K.J, Rebecca, R.A, Ysidor, K.N. (2019b) Development of a method to produce granulated sugar from the inflorescences sap of coconut (*cocos nucifera* l.) in Ivory Coast: Case of Hybrid PB113+. *J Exp Agric Int* 39(2):1-9. 10.9734/JEAI/2019/v39i230331.

Muriel, J.O.D., Louis, K.K.J., Rebecca, A.R. (2020b) Sensory properties of table sugars derived from the inflorescences sap of three coconut (*Cocos nucifera* L) cultivars in Côte d'Ivoire. *Am J Food and Nutrn* 8(3): 90-100. doi: 0.12691/ajfn-8-3-6.

Muriel, O.D.J., Rebecca, A.R., Louis, K.K.J. (2021) Mineral composition of table sugars from the sap of the inflorescences of 03 coconut (Cocos *Nucifera* L.) cultivars in Ivory Coast. *J Chem Biol Physical Sci* 11(1):82-90, https://doi.org/10.24214/jcbps.a.11.1.08290.

Muriel, O.D.J., Louis, K.K.J., Rebecca, A.R. (2020a) Functional properties of table sugars derived from the sap of the inflorescences of 03 coconut (*Cocos nucifera* L.) cultivars in Côte D'Ivoire. *Int J Adv Res* 8(11):254-263, doi: 10.21474/IJAR01/12004.

Muriel, O.D.J., Rebecca, R. A., Louis, K.K.J., Ysidor, K.N. (2019a) Physico-chemical characterization of granulated sugar from coconut (*cocos nucifera* l.) inflorescence sap cultivars and sugar cane in Côte d'Ivoire. *Current J Applied Sci Technol* 37(3): 1-13, doi: 10.9734/CJAST/2019/v37i330287.

Nagai, T., Kai, N., Tanoue, Y., Suzuki, N. (2018). Chemical properties of commercially available honey species and the functional properties of caramelization and Maillard reaction products derived from these honey species. *J Food Sci Technol* 55(2):586-597, https://doi.org/10.1007/s13197-017-2968-y.

Nakamura, S., Watanabe, A., Chongpraditnun, P., Suzui, N., Hayashi, H., Hattori, H., Chino, M(2004) Analysis of phloem exudate collected from fruit-bearing stems of coconut palm: palm trees as a source of molecules circulating in sieve tubes. *Soil Sci Plant Nutr* 50(5): 739-745.

Naknean, P., Meenune, M., Roudaut, G. (2010) Characterization of palm sap harvested in Songkhla province, Southern Thailand. *Int Food Res* J 17:977–986.

Naknean, P., Meenune, M., Roudaut, G. (2013). Changes in properties of palm sugar syrup produced by an open pan and a vacuum evaporator during storage. *Inter Food Res J* 20(5):2323-2334.

Naufalin, R., Tri, U.Y., and S. Anna S. (2013) Types and concentration effect of natural preservatives on the quality of palm sugar. *Teknol. Pertan* 14(3)165–174

Njugu, SP(2013) Competitiveness of coconut wine value chain in the coconut industry in kenya. Developing Country Studies 3(6): 108-114.

Novarianto, H., Tulalo, M.A., Mawardi, S. (2021a) Preliminary investigation of the potential of coconut sugar production using dwarf varieties. *CORD* 37:23-32, doi: https://doi.org/10.37833/cord.v37i.444.

Novarianto, H., Tulalo, M.A., Mawardi, S. (2021b) Coconut sugar production of dwarf coconut varieties. *IOP Conf. Series: Earth and Environmental Science* 800, 012024, doi:10.1088/1755-1315/800/1/012024, 10 pages.

Nurhadi, B., Roos, Y.H. (2017). Influence of anti-caking agent on the water sorption isotherm and flow-ability properties of vacuum dried honey powder. *J Food Eng* 210: 76–82.

Nurhadi, B., Sukri, N., Saputra, R.A., Wandhani, F.I., Nurlita, A.I. (2020) *Physical characteristics of amorphous and crystalline coconut sugar powder with the addition of tricalcium phosphate (TCP) as an anticaking agent.* 2020, 5320173, https://doi.org/10.1155/2020/5320173, 10 pages.

Nurhadi, B., Sukri, N., Sugandi, W.K., Widanti, A.P., Restiani, R., Noflianrini, Z., Rezaharsamto, B., Herudiyanto, M. (2018) Comparison of crystallized coconut sugar produced by traditional method and amorphous coconut sugar formed by two drying methods: vacuum drying and spray drying. *Int J Food Properties* 21(1):2339–2354, https://doi.org/10.1080/10942912.2018.1517781.

Nusaa, C.P., Rimbawan, R. (2022) Macronutrient and sensory analysis of block, crystal, and liquid coconut sugar. *Bulletin of Culinary Art and Hospitality* 2(2):51-56

Okoma, D.M.J., Konan, K.J.L., Assa, R.R. (2020). Vitamin and osidic composition of table sugars from the inflorescences sap of 03 coconut cultivars (*Cocos nucifera* L.) in Ivory Coast. *Food and Nutn Sci* 11 (12): 1117–1126.

Phaeon, N., Chapanya, P., Mueangmontri, R., Pattamasuwan, A., Lipan, L., Carbonell-Barrachina, Á.A., Sriroth, K, Nitayapat, N (2021). Acrylamide in non-centrifugal sugars and syrups. *J Sci Food Agric* 101:4561–4569.

Philippe, R.N., De Mey, M., Anderson, J, Ajikumar, PK (2014). Biotechnological production of natural zero-calorie sweeteners. *Current Opinion in Biotechnol* 26:155–161.

Prakobsil, J., Benjabong, W., Nitithamyong, A., Tavichatwitayatkul, R., Karnpanit, W., Moungsrichan, N., Kingsoong, K. (2010). Assessment of sanitation of traditional coconut sugar in Samut Songkram Province, Thailand. *Toxicol Letters* 196S:P309-044, doi: 10.1016/j.toxlet.2010.03.1078.

Pratama, F., Susanto, W. H. and Purwantiningrum, I. 2015 Making coconut sugar from natural fermented sap. *Jurnal Pangan and Agro Industri* 3(4):1272-82.

Provost, J. (2019). The Maillard Reaction. In: *Food Aroma Evolution.* CRC Press, pp 281–291.

Psomiadis, D., Zisi, N., Koger, C., Horvath, B., Bodiselitsch, B. (2018). Sugar-specific carbon isotope ratio analysis of coconut waters for authentication purposes. *J Food Sci Technol* 55(8):2994–3000, 10.1007/s13197-018-3217-8.

Purnomo, H. (2007). Volatile components of coconut fresh sap, sap syrup and coconut sugar. *ASEAN Food J* 14 (1): 45.

Purnomo, H., Suryoseputro, S. (2001). Traditional coconut sugar production in indonesia. a comparative study on the technology and physico-chemical properties of coconut sugar from four villages in East Jawa and Bali. In: *The 11th Biennial International Congress of Asian Regional Association for Home Economics* (ARAHE). pp 1–6.

Puspitasari-Nienaber, S.B., Sumaryanto, H. (1996). Analysis of volatiles of kecap manis (a typical Indonesian soy sauce). *Flavour Science: Recent Developments*, 197, 62.

Putra, A.P., Romadhona, K.N., Anggraeni, R.S., Julyanti, A.R., Widiastuti, R.F., Diniatik (2019). Guava leaves, corn cobs, and mangosteen peels as antioxidants and national standardized natural preservatives for coconut brown sugar. Asian *J Pharm Clin Res* 12(9):1-4, doi: http://dx.doi.org/10.22159/ajpcr.2019.v12i9.29216.

Rakphon, A., Srijesadaruk, V. (2017). Effect of thermal processing on physical, chemical properties and volatile compounds of coconut (*Cocos nucifera* L.) sugar. In *Proceedings of the 4th International Postgraduate Symposium on Food, Agriculture and Biotechnology (IPSFAB),* Maha Sarakham, Thailand, 30–31 August 2017; pp. 80–91.

Ramya, H.N., Gowda, M.C., Jayamala, G.B., Krishnamma, P.N., Shreedevi, M.S. (2010). Effect of ambient atmospheric conditions on change in physico- chemical properties of stored jaggery. *Int J Applied Agri Res* 5(1):9-18.

Rasmussen, H., Sørensen, H.R., Meyer, A.S. (2014). Formation of degradation compounds from lignocellulosic biomass in the biorefinery: Sugar reaction mechanisms. *Carbohydrate Res* 385: 45–57. https://doi.org/10.1016/j.carres.2013.08.029.

Renjith, R.S., Chikku, A.M., Rajamohan, T. (2013). Cytoprotective, antihyperglycemic and phytochemical properties of *Cocos nucifera* (L.) inflorescence. *Asian Pacific J Tropical Med* 6 (10): 804–810.

Rogers, K.M., Phillips, A., Fitzgerald, J., Rogers, P., Ferguson, C., Cooper, J., Yuan, Y. (2021) Authentication of indonesian coconut sugar using stable carbon isotopes. Food Anal Methods https://doi.org/10.1007/s12161-021-01967-9, 6 pages.

Roman, S., Sánchez-Siles, L.M., Siegrist, M. (2017). The importance of food naturalness for consumers: Results of a systematic review. *Trends in Food Sci Technol* 67:44–57.

Roos, Y.H., Drusch, S. (2015). Phase transitions in foods. *Academic Press*. 360p.

Salil, G., Nevin, K.G., Rajamohan, T. (2011). Arginine rich coconut kernel protein modulates diabetes in alloxan treated rats. *Chemico-Biological Interactions* 189 (1–2): 107–111.

Samarajeewa, U., Wijeratne, M.C.P. (1979) Methods for determining suitability of coconut sap for preparation of jaggery, sugar and golden syrup. *Ceylon Coconut Quart* 30:72-80.

Saputro, A.D., Van de Walle, D., Dewettinck, K. (2019). Palm sap sugar: a review. *Sugar Tech*, 21: 862–867.

Saraiva A., Carrascosa C., Raheem D., Ramos F., Raposo A. (2020). Natural sweeteners: The relevance of food naturalness for consumers, food security aspects, sustainability and health impacts. *Int J Environ Res Public Health* 17 (17):6285.

Saraiva, A., Carrascosa, C., Ramos, F., Raheem, D., Lopes, M., Raposo, A. (2023). Coconut sugar: chemical analysis and nutritional profile; health impacts; safety and quality control; food industry applications. *Int J Environ Res Public Health* 20 (4): 3671.

Sawitzki, F., da Silva, M.A.M. (2023) Glycemic response of coconut sugar, sucrose and brown sugar in healthy subjects. *Uniting knowledge integrated scientific research for global development,* 10.56238/uniknowindevolp-014, 10 pages.

Sekhon, B.S. (2009) Pharmaceutical Co-Crystals – A Review. *Ars Pharm* 50(3): 99–117.

Shetty, S.S., Ramesh, S.V., Arivalagan, M., Roopashree, P.G., Manikantan, M.R., Hebbar, K.B., Kumari, S. (2023). Coconut (*Cocos nucifera* L.) inflorescence sap-derived sugar restores the glucose and lipid homeostasis in streptozotocin-induced diabetic Wistar rat model. *Biomed* 43 (3):859–866.

Sohn, M., Ho, C.T. (1995). Ammonia generation during thermal degradation of amino acids. *J Agric Food Chem* 43 (12):3001–3003.

Srikaeo, K., Thongta, R. (2015). Effects of sugarcane, palm sugar, coconut sugar and sorbitol on starch digestibility and physicochemical properties of wheat based foods. *Int Food Res J* 22 (3): 923.

Starzak, M., Mathlouthi, M. (2010). Formation of amorphous sugar in the syrup film–a key factor in modelling of industrial sugar drying. *Food Chem* 122 (2):394–409.

Sulistyo, S.B., Sudarmaji, A., Haryanti, P., Kuncoro, P.H. (2024) A novel approach for detection of granulated coconut sugar adulteration using LED-based spectrometer and machine learning, *Information Processing in Agriculture* (2024), doi: https://doi.org/10.1016/j.inpa.2024.09.007.

Swain, S., Din, M., Ram, C., Gangaiah, B. (2016). Post harvest processing and value addition in coconut, arecanut and spices in Islands. Gangaiah B, Roy SD(eds), *Integrated Farming Systems for Tropical Islands of India*, pp 152–157.

Swasti Y.R., Purwijantiningsih F., Pranata F. (2024) Antioxidant activities and mutagenic compounds in coconut and palm sugar from special region of Yogyakarta Indonesia. *Food Res* 8 (4) : 336 – 342, doi: 10.26656/fr.2017.8(4).251.

Tiarasari, E., Wulandari, R., Rahayoe, S., Amanah, H.Z. (2023) Drying profile of coarse coconut palm sap sugar using simple air-oven dryer Conference: *International Conference on Sustainable Environment, Agriculture and Tourism (ICOSEAT 2022),* doi: 10.2991/978-94-6463-086-2_60 pp 443-447.

Tremblay P., Paquin R. (2007) Improved detection of sugar addition to maple syrup using malic acid as internal standard and in 13C isotope ratio mass spectrometry (IRMS). *J Agric Food Chem* 55(2):197–203, doi: 10.1021/jf062413a.

Trinidad, T., Mallillin, A., Avena, E., Rodriguez, R., Borlagdan, M., Cid, K.B., Biona, K. (2015). Coconut sap sugar and syrup: A promising functional food/ingredient. *Acta Manila. Ser. A* 63:25–32.

Trinidad, T.P. (2015) Characterizing coconut sap sugar and syrup as a promising functional food/ingredient. *BMJ* Open 2015;5(Suppl 1):A33.

Trinidad, T.P., Mallillin, A.C., Sagum, R.S., Encabo, R.R. (2010). Glycemic index of commonly consumed carbohydrate foods in the Philippines. *J Functional Foods* 2 (4):271–274.

Tulalo, M.A., Mawardi, S. (2018) Potensi Produksi Nira Dan Gula Tiga Aksesi Kelapa Genjah (Potential sap and coconut sugar production of three accession dwarf coconut) . *Industrial Crops Res J* 24(2):87-92, doi: 10.21082/littri.v24n2.2018.87-92.

Uppal, S., Sharma, S. (1999). Evaluation of different methods of jaggery (gur) storage in subtropical region. *Indian J of Suga Techn.* 14(1):17-21.

Weerawatanakorn, M., Asikin, Y., Kamchonemenukool, S., Tamaki, H., Takara, K., Wada K. (2021). Physicochemical, antioxidant, volatile component, and mass spectrometry-based electronic nose analyses differentiated unrefined non-centrifugal cane, palm, and coconut sugars. *J Food Meas Charact* 15:1563–1577.

Wiboonsirikul, J. (2016). Differentiation between adulterated and non-adulterated palm sap using physical and chemical properties combined with discriminant analysis. *Int Food Res J* 23(1):61–67.

Wijewickreme, A.N., Krejpcio, Z., Kitts, D.D. (1999) Hydroxyl scavenging activity of glucose, fructose, and ribose-lysine model Maillard products. *J Food Sci* 64: 457-461.

Witham, K. (2016) *Coconut sugar: Germany's next trendy sugar alternative?* https://www.mintel.com/insights/ingredients-flavours-and-fragrances.

Wrage, J., Burmester, S., Kuballa, J., Rohn, S. (2019). Coconut sugar (*Cocos nucifera* L.): Production process, chemical characterization, and sensory properties. *LWT*, 112: 108227.

Xia, Q., Li, R., Zhao, S., Chen, W., Chen, H., Xin, B., Huang, Y., Tan, M. (2011). Chemical composition changes of post-harvest coconut inflorescence sap during natural fermentation. *African J Biotechnol* 10(66):1499915005. https://doi.org/10.5897/AJB10.2602.

Zdiniakova, T., Calle, M.B. (2020). Feasibility study about the use of element profiles determined by ED-XRF as screening method to authenticate coconut sugar commercially available. *Eur Food Res Technol* 246:2101–2109, https://doi.org/10.1007/s00217-020-03559-z.

Zulfia, V., Ainuri, M., Khuriyati, N., Yusuf, R., Alim, A.S., Pato, U. (2022) Optimizing of the parameters of coconut sugar production using taguchi design in Riau, Indonesia. *Int J Adv Sci Eng Infor Tech* 12(2):752-758, doi: 10.18517/ijaseit.12.2.15684.

Zulfia, V., Ainuri, M., Khuriyati, N. (2019) Modifikasi Parameter Produksi untuk Meningkatkan Mutu Kimia Gula Kelapa Cetak di Kabupaten Indragiri Hilir, Provinsi Riau (Modification of production parameters to improve coconut sugar quality in Indragiri Hilir District, Riau Province). *Jurnal Teknologi dan Manajemen Agroindustri* 8(3):197-208, doi:10.21776/ub.industria.2019.008.03.4.

Zuliana, C., Widyastuti, E., Susanto, W.H. (2016). Pembuatan gula semut kelapa (kajian ph gula kelapa dan konsentrasi natrium bikarbonat). *Jurnal Pangan dan Agroindustri* 4(1): 109-119.

Chapter 9

The Health and Nutritive Potential of Coconut Palm Sap and Value-Added Products

S. V. Ramesh[1]
V. Krishnakumar[2]
and K. B. Hebbar[1,*]
[1]ICAR-Central Plantation Crops Research Institute, Kasaragod, Kerala, India
[2]ICAR-Central Plantation Crops Research Institute, Kayamkulam, Kerala, India

Abstract

Coconut palm sap, or *neera*, is a nutritious beverage tapped from the inflorescences of the coconut palm. As a phloem sap, *neera* is rich in nutrients such as sugars, vitamins, ascorbic acid, phenolics, and has significant antioxidant potential. *Neera* has been processed into a variety of value-added products, including coconut sugar and other secondary products. Given its nutritional potential, low glycemic index, and rich mineral content, coconut sap and its derivatives have been extensively studied in the context of human health and nutrition. Research has demonstrated that coconut *neera* and sugar have several health benefits, including anti-diabetic and hypoglycemic effects, reduction of hypertension, antioxidant activity, cardiovascular protection, nephroprotective action, and digestive health improvements. This chapter explores the potential and proven health benefits of coconut *neera* and sugar in light of recent scientific evidence.

Keywords: coconut sugar, hypoglycemia, hepatoprotective effect, nephroprotective effect, health benefits

---

* Corresponding Author's Email: hebbar.kb@icar.gov.in

In: Coconut Palm Sap
Editors: K. B. Hebbar, V. Krishnakumar and S. V. Ramesh
ISBN: 979-8-89530-640-6

1. Introduction

The coconut palm, a staple in South Asia, is renowned for its exceptional nutritional and medicinal properties. This versatile palm offers a multitude of health benefits and sustains human needs. Thriving in diverse environments, coconut palms contribute to ecological balance by improving soil health and requiring minimal water. As one of the world's oldest cultivated plants, this palm has been a cornerstone of many cultures for centuries. Coconut trees produce flowers or inflorescences year-round, allowing for the practice of tapping them for sap. Coconut sap, also known as *neera*, is a sweet liquid extracted from the unopened flower cluster of a coconut palm. This traditional method has been used for centuries in tropical regions. The nutrient-rich sap, collected before the inflorescence turn into coconuts, is a versatile beverage and ingredient (Hebbar et al., 2018; Misra 2016). A single coconut tree can produce multiple flower clusters each year, with potential to yield numerous coconuts. However, tapping an inflorescence cluster for sap can produce significantly more liquid compared to allowing it to develop coconuts. The sap, known as *neera*, *raa*, or *tuba* in different regions, is naturally sweet and packed with essential nutrients. It can be fermented into toddy or processed to preserve its nutritional value. Consumed fresh, it is prized for its taste and health benefits. Tapping a coconut tree can provide a consistent sap flow for up to 20 years and is more sustainable in terms of energy and resource use compared to coconut oil production (Hebbar et al., 2015a). This traditional practice, common in coconut-growing regions, involves carefully cutting the flower cluster to collect the sap. To maintain its quality, sap extraction is typically done early in the morning as it ferments quickly at normal temperatures. The collected sap is stored in earthenware containers to prevent fermentation, resulting in a sweet, nutritious, and non-alcoholic beverage.

2. Processing of Coconut Inflorescence Sap

A coconut tree typically produces 10 to 14 spadices year^{-1}, depending on its health and the local climate. Each spadix can yield approximately 2 l of sap per day, and a single coconut tree can produce around 400 l of sap annually. If the sap remains unfermented, it can be sold as a nutritious beverage or concentrated into products such as syrup, jaggery, and brown sugar (Hebbar et al., 2022; Thampan 2013). Alternatively, the fermented sap can be used to make toddy or wine. The adventof the 'Cocosap chiller' (Hebbar et al., 2015b; Hebbar and Augustine 2021) has significantly addressed the challenges associated with collecting hygienic and unfermented coconut sap. This innovation has also contributed to the easing of the ban on sap tapping in many coconut-growing regions. As a result, licenses are now being issued for sap tapping, processing, and marketing, either to individual entrepreneurs or farmer organizations such as Coconut Producer Companies (CPCs) and Coconut Producer Federations (CPFs) in Indian states like Goa, Karnataka, Kerala, Tamil Nadu, and Telangana. Some of these entrepreneurs and organizations have begun producing coconut sugar using advanced scientific methods as a commercial enterprise. Due to its high nutritional quality and low glycemic index (GI), palm sugar is considered a premium product and commands relatively high prices in both the domestic and international markets.

Palm sap, (also known as *neera* or *Kalparasa*), is rich in sugar, typically around 14–16%. When heated, the moisture in the sap evaporates, leading to the crystallization of sugar (Hebbar et al., 2015a; Wrage et al., 2019). Unfermented sap with this sugar content crystallizes easily upon heating without the need for additional chemicals. For optimal sugar production, the sap should have a pH of around 6.7; lower pH levels result in a semisolid gum-like texture or hard lumps upon boiling. To prevent these issues, traditional methods for making sugar or jaggery often involve adding substances such as lime, plant extracts, oils, or mucilaginous materials like *Abelmoschus* species and castor seeds. During heating, sugars in the palm sap undergo Maillard reactions and caramelization, which contribute to the sweet, toasty, and nutty caramel-like aroma of the sugar. Volatile compounds like pyrazines, furans, ketones, fatty acids, and organic acids are responsible for these aromas. Pyrazines, for example, are released when the sap is heated above 110°C (Babasaheb et al., 2007).

The quality of *neera*, as well as the boiling temperature and heating time influences the volatile components of the palm sap (Martins et al., 2001). Higher sugar yields and better quality are achieved with high-quality sap (pH around 7 and 14 to 16° Brix), while yields decline with lower-quality sap. Various processing methods are employed depending on the type of heating and moisture extraction methodology used. The unfermented sap derived from the tender, unopened inflorescence of coconut trees is recognized for its significant health benefits. Due to concerns about thermal degradation associated with the traditional evaporation process for producing coconut sugar, a low-temperature method of coconut sugar production was developed

[Coconut Sap Powder, registered as COCOZEN™] (Jose et al., 2017; Ratheesh et al., 2017). This method not only preserves the micronutrients in the coconut sugar, but also maintains the bioactive molecules that provide multiple health benefits, including supporting normal kidney and liver function, as well as offering antioxidant and anti-inflammatory properties.

The collected sap undergoes several stages of processing, including filtration, refrigeration, centrifugation, and packaging. This treated sap can be stored up to two months without the loss of its quality features. The Defense Research and Development Organization (DRDO) of India, in collaboration with the Central Food Technology Research Institute (CFTRI), have developed a method to preserve coconut sap (*neera*) without it fermenting into alcohol (toddy). The process involves collecting the sap from unopened inflorescence, chilling it immediately to 4°C, followed by filtration and further chilling to 2–8°C. Citric acid (0.04–0.2%) and nisin (10–15 ppm) are then added, and the sap is pasteurized at 95°C for 5–10 min for long-term preservation. Centrifugation at 4000 rpm for 10 min removes suspended particles. The heat-treated *neera* remains stable for up to 72 h. A shelf life of one year is achievable with in-pack pasteurization or retort pouch processing when stored in refrigerated conditions or for 30 days at room temperature. However, undesirable taste and odour can develop, making it unfit for consumption. Due to the photosensitivity of unfermented coconut sap, which contains ascorbic acid, storage in polyethylene terephthalate (PET)/aluminum foil packaging was also developed. Another process was created to preserve *neera* by maintaining its cloudiness and natural components while deodorizing it. This method includes collecting the sap, filtration, chilling to 2–8°C, and adding bentonite or activated carbon black. The mixture is stirred, centrifuged, and filtered before pasteurization or the addition of preservatives to produce *neera* with its natural physical featuresand reduced odour. Additionally, value-added products like palm jaggery, palm syrup, and palm sugar, *neera* based sweets, ice cream can be made from this unfermented sap (Asha et al., 2019; Beegum et al., 2022; Ghosh et al., 2016).

3. Nutritional Features of Coconut Inflorescence Sap

Unfermented coconut sap, extracted from immature inflorescences, varies in colour from light yellow to cream to dark brown, with a sweet taste and a pleasant nutty aroma (Borse et al., 2007). It is free from impurities or extraneous matter. Fresh sap typically has a basic pH, though this can vary slightly between trees. The sap begins to ferment within 2–3 h at room temperature, during which the pH gradually declines, reaching around 3.5 when fully fermented. When stored in a deep freezer (at -2°C to -1°C), the sap remains fresh, and no significant change in pH is observed. Fresh sap contains approximately 15% sugars, which decrease to about 6% at a pH of 4. The sugar content further decreases to 5% as pH drops.

When exposed to air, fresh sap undergoes initial lactic acid fermentation followed by alcoholic fermentation due to microbial activity (Gopal et al., 2021). The sap is a good source of nutrients that support a healthy diet. A 100 ml serving provides 75 calories and 250 mg of proteins, along with essential minerals, vitamins, and amino acids. It is naturally rich in magnesium, zinc, potassium, and iron and contains around 16% carbohydrates, with sucrose being the primary component, along with small amounts of glucose, fructose, inositol, and raffinose (Hebbar et al., 2020). Fresh sap also contains 16 amino acids, with glutamic acid, threonine, and aspartic acid being the most abundant, and ascorbic acid as a key vitamin. Ascorbic acid is a powerful dietary antioxidant that helps mitigate the harmful effects of reactive oxygen and nitrogen species, which can cause oxidative damage to macromolecules like lipids, DNA, and proteins. This oxidative damage is linked to chronic diseases such as cardiovascular disease (CVD), stroke, cancer, neurodegenerative diseases, and cataract genesis.

Fresh coconut sap is particularly high in glutamic acid and aspartic acid, two amino acids naturally produced by the body. These amino acids offer numerous health benefits, especially for cognitive function and metabolism, and serve as precursors to other amino acids. Unfermented sap contains 84% moisture and 0.04% ether extractives (volatiles), which include ethyl lactate, 3-hydroxy-2-pentanone, phenyl ethyl alcohol, 2-methyl tetrahydrofuran, farnesol, and tetradecanone. The astringency and harsh notes in fermented *neera* are likely due to increased levels of acids (19.0 mg l^{-1}), such as dodecanoic acid and palmitoleic acid, along with higher concentrations of ethyl alcohol and ethyl esters. Fresh coconut sap helps keep the body hydrated while its nutrients nourish and energize, leaving one feeling refreshed.

4. Health Benefits of Coconut Inflorescence Sap

Coconut inflorescence sap is a flavourful drink that is becoming increasingly popular due to its high nutritional value. It has a soothing and cooling effect and is celebrated in traditional medicinal systems for its rejuvenating properties. It is traditionally recommended for treating anemia, tuberculosis, bronchial congestion, and piles (Hali 2013). Coconut inflorescence sap is valued for its high levels of glutamic acid, an important amino acid, which serves as a precursor for protein production. Enriched with different vitamins and electrolytes, the sap also acts as a source of coenzymes and cofactors in various metabolic processes, thereby offering protection against diseases and enhancing overall immunity (Jose et al., 2017). The rich mineral content of coconut inflorescence sap plays a critical role in maintaining the body's electrolyte balance, which is essential for avoiding or reducing dehydration, helps regulate fluid balance within cells and tissues and contributes to maintaining normal body temperature. It is recognized as a potent natural beverage for preserving body fluids and is also deemed beneficial as a post-operative drink (Asha et al., 2019).

Coconut sugar crystals, made from this pure, natural sap, are a low-glycemic sweetener. Unlike most brown sugar, which is boiled at temperatures up to 221°F and ends up containing 93% sucrose, coconut sap crystals contain only 0.5% glucose, 1.5% fructose, 16% sucrose, and 82% inulin - a prebiotic that promotes digestive health (Hebbar et al., 2020; Hebbar et al., 2022). These sap crystals serve as an ideal sweetener. The major health benefits of fresh, sweet coconut sap include its nutritional richness and its potential to support overall well-being.

4.1. Anti-Diabetic Activity

The young inflorescence of the coconut palm has demonstrated protective and restorative effects against alloxan-induced pancreatic cytotoxicity and severe hyperglycemia. These benefits are achieved by enhancing peripheral glucose utilization, improving impaired liver glycolysis, inhibiting gluconeogenesis, and repairing and restoring the remaining beta cell population in the pancreas. These effects are likely due to the presence of phenolic acids, flavonoids, and other phytochemicals, which may work synergistically or independently to modulate glycolytic and gluconeogenic enzyme activity. The findings of this study provide scientific support for the use of coconut inflorescence as a promising candidate in traditional medicine for the treatment of diabetes (Raveendran and Thankappan 2012).

The low GI of coconut sap helps in slow uptake of glucose in the bloodstream, thereby contributing to the fight against diabetes by reducing insulin requirements. Alloxan-induced cytotoxicity can damage these cells, leading to decreased insulin production and glucose intolerance. The results of the studies on the effects of pre and post-treatment of young coconut inflorescence sap on alloxan-induced diabetic rats indicated that sap may possess potential properties to promote the regeneration or repair of damaged islet cells, restore their function and enhance insulin secretion (Renjith and Rajamohan 2012a; 2012b). The non-toxic nature and the cytoprotective and antihyperglycemic properties of coconut inflorescence sap was also reported (Renjith et al., 2013). Coconut sap contains nicotinic acid and riboflavin and can be consumed by people suffering from diabetes since it has a low GI and a low glycemic load (Femi et al., 2020). Wistar rats with streptozotocin-induced diabetes were administered with coconut palm sugar. This resulted in bringing back carbohydrate and lipid metabolism to normal levels in such rats. The oral administration of coconut sugar to Wistar rats with streptozotocin-induced diabetes has shown several health benefits. The study observed positive outcomes, particularly in normalizing carbohydrate and lipid metabolism in diabetic rats. Additionally, the use of coconut sugar produced favorable results in reducing oxidative stress associated with diabetes (Shetty et al., 2023).

The GI measures how quickly and how much a specific carbohydrate raises blood sugar levels by releasing glucose into the bloodstream. Glycemic load, which factors in both the quality and quantity of carbohydrates, is another useful measure for managing blood sugar levels. A low glycemic load diet is particularly helpful for diabetes and those managing their body weight, as it reduces the overall impact of carbohydrates on blood sugar levels. The glycemic load of coconut palm sugar is 1.4, or rounded down to 1, indicating it has a minimal impact on blood sugar. Research by the Food and Nutrition Research Institute of India (FNRI) in 2007 found that coconut sugar has a naturally low GI of 35, which is significantly lower than most commercial sugars—table sugar has a GI of 70, honey has 55, and cane sugar has 68. Replacing high-GI sugars with low-GI alternatives like coconut sugar can help reduce blood glucose levels in individuals with diabetes. Studies have also shown a link between high insulin levels and certain cancers, such as pancreatic, colorectal, prostate, and breast cancer. Moreover, diets high in sugar, refined carbohydrates, and glycemic load have been associated with an increased risk of cancer. It is recommended that lifestyle changes, including maintaining a healthy weight, exercising regularly, and following a low-GI diet, may help protect against cancer, at least partially, by lowering insulin levels.

4.2. Anti-Hypertension

Hypertension or high blood pressure is a major risk factor for conditions such as stroke, kidney damage, and cardiovascular diseases (CVDs). Coconut inflorescence sap has been found to be effective in managing hypertension and its associated complications when taken as a supplement (100 ml day^{-1}) without any side effects. This antihypertensive effect is thought to be due to the presence of active biological compounds in the sap, which help normalize calcium channels and the nitric oxide pathway. This normalization provides anti-atherogenic effects, improves endothelial dysfunction, and reduces peripheral resistance, all of which contribute to lowering blood pressure. Consumption of 100 ml of coconut sap day^{-1} has proven effective in managing hypertension and its associated complications, with no observed side effects. This efficacy is attributed to the active biological constituents in coconut *neera*, which offer a range of therapeutic benefits, including antioxidant, hypolipidemic, hypoglycemic, diuretic, and antihypertensive effects.

Panonnummal et al., (2021) coined the term as non-alcoholic palm nectar (NPNC) for the sap obtained from the juvenile inflorescence of coconut palm. Due to the physicochemical parameters, nutritional composition, antioxidant potential, and therapeutic health benefits, NPNC can be a better alternative for the commercially available unhealthy carbonated beverages and energy drinks. NPNC and the two other products, i.e., honey and sugar exhibited stimulatory effect on humoral and cell-mediated immunity. NPNC presented significant diuretic effect and saluretic effect. Due to the abundance of iron in coconut sap-based honey, it might be a good alternative with better medicinal value over the normal honey.

The antihypertensive effects of coconut *neera* are achieved through the normalization of nitric oxide (NO) pathways and calcium channels. This contributes to its anti-atherogenic properties, improves endothelial function, and enhances peripheral resistance, ultimately leading to reduced blood pressure. Furthermore, the sap contains free amino acids, including sulfur-containing amino acids (such as cysteine, L-arginine, and tyrosine), as well as antioxidants, vitamins A and C. Intake of 100 ml fresh sap for five subsequent weeks in the morning by hypertensive women reduced their systolic and diastolic blood pressure as well as serum total cholesterol (Bhagya and Soumya 2016), which could be due to the presence of biologically active constituents in coconut inflorescence sap possessing hypocholesterolaemic and antioxidant properties thereby reducing the blood pressure.

4.3. Antioxidant Activity

The antioxidant properties of coconut inflorescence sap have been extensively studied, focusing on its reducing power, ascorbic acid levels, polyphenol content, and alpha-amylase inhibitory activity. Polyphenols and ascorbic acid are potent antioxidants with strong free radical scavenging abilities, which contribute to the sap's significant reducing power. Alpha-amylase inhibitors present in the sap slow down the digestion of carbohydrates, thereby reducing the rate of glucose absorption. Antioxidants have been shown to inhibit cancer initiation caused by various chemical carcinogens and radiation across multiple animal studies, including those conducted on mice, rats, and hamsters. Epidemiological research suggests that a diet rich in plant-based products containing natural antioxidants can reduce the risk of cancer. Vitamin C, or L-ascorbic acid, is known to effectively neutralize superoxide, hydrogen peroxide, hydroxyl radicals, peroxyl radicals, and singlet oxygen (Valko et al., 2006). Ascorbic acid also protects cell membranes from peroxidation by enhancing the activity of tocopherol, the primary lipid-soluble vitamin, and is considered the most important antioxidant in extracellular fluids. In traditional medicine, *neera* is used to treat conditions such as bronchial suffocation, anemia, tuberculosis, and piles. An *in vitro* analysis of their free radical scavenging properties revealed that *neera* has a higher antioxidant activity compared to toddy. When tested on mice, *neera* demonstrated a stronger ability to mitigate the oxidative damage caused by sodium fluoride. *Neera* effectively reduces metal ions and scavenges hydroxyl and DPPH radicals, with IC50 values of 6.5 ± 1.9 µl ml^{-1} and 44 ± 3.14 µl ml^{-1}, respectively (Silpa et al., 2021).

Coconut inflorescence sap, being rich in ascorbic acid, exhibits strong antioxidant capacity, which may be beneficial in combating oxidative stress associated with cancer therapy. The study highlighted the antioxidant benefits of *neera*, a non-alcoholic natural beverage, and suggests it as a healthier alternative to synthetic carbonated soft drinks. Regular consumption of *neera* may help protect the body from various chronic diseases, particularly those where oxidative stress is a significant factor (Devi et al., 2015). Fresh sap was found to scavenge superoxide anion, hydroxyl radical and H_2O_2 better than the naturally fermented sap; in contrast, the latter showed a more powerful ability to prevent damage to DNA than the fresh sap (Chen et al., 2011). The naturally fermented sap had a higher polyphenol level but lower ascorbic acid content than fresh sap, which suggested that fermentation is beneficial for improving the polyphenol content of coconut sap. Prabha et al., (2017) reported that the high phenolic content (176.45 μg mg^{-1} GAE) of coconut inflorescence sap has scavenging activity against several free radicals. Further, fresh inflorescence sap exhibited high antioxidant activity than its fermented form. There was considerable increase in SOD, CAT and GPx activities and GSH level in mice liver due to the supplementation of fresh sap for two weeks (Prabha et al., 2021). Sulistyo and Haryanti (2020) showed that coconut sap added by basic amino acid i.e., 0.75 mM lysine produced a significant increase of total phenolic content and radical scavenging activity in the sap until the heating temperature reached 118°C.

The antioxidant activity of the sap tapped in sunny weather condition with the use of mangosteen peel powder at 0.84 g l^{-1} was 34.39% radical scavenging activity (RSA) (Haryanti et al., 2018). Arginine contains positive charged ends that are suitable to bind molecules with negative charge like carbonyl groups of reducing sugar in coconut sap. Haryanti et al., (2021) found that a concentration of 0.8 mM of arginine and the sap temperature of 100°C were the optimum condition to obtain the highest DPPH radical scavenging activity i.e., 69.93%, whereas, the chelating activity of coconut sap added with 0.4 mM was significantly higher than other treatments (34.42%).

Coconut inflorescence sap contains significant amount of micro and macronutrients. The phytochemical analysis of coconut inflorescence sap showed the presence of flavonoids, alkaloids, terpenes, steroids, phenols, and glycosides. Detailed analysis indicated the presence of biologically significant compounds, namely 4-hydroxy coumarin, p-coumaric acid, mellein, leucopelargonidin, coumarin, 3 caffeoylquinic acid, ferulic acid 4-O-glucosidase and also three essential amino acids L-phenylalanine, L-isoleucine and D-tryptophan (Sharafudeen et al., 2022). Coconut inflorescence sap has a significant beneficial effect in upgrading the antioxidant status. Aheeshan et al., (2022) also reported about the antioxidant activity of sap.

4.4. Effect on Cardiovascular Disease (CVD)

Atherosclerosis, a chronic disease that begins with risk factors leading to the development of subclinical conditions and eventually progressing over to cardiovascular disease (CVD), is significantly influenced by oxidized low-density lipoprotein (OxLDL). OxLDL plays a critical role in the development of atherosclerosis by promoting inflammation and lipid accumulation in the arterial walls. Recent studies have explored the anti-inflammatory effects of coconut inflorescence sap, particularly in the form of coconut sap powder, against OxLDL-induced inflammatory responses in human peripheral blood mononuclear cells (hPBMCs). Coconut inflorescence sap can down regulate and reverse OxLDL-induced changes, suggesting its potential anti-inflammatory effects through the TLR-NF-κB signaling pathway (Ratheesh et al., 2017).

Similar to type 2 diabetes, diets high in refined and high GI carbohydrates have been linked to an increased risk of heart disease. These foods elevate blood insulin levels, leading to higher blood fats (triglycerides), lower levels of high-density lipoprotein (HDL or 'good' cholesterol), increased blood pressure, and a greater tendency for dangerous blood clots to form and persist. Ascorbic acid, a key component of coconut inflorescence sap, is known to prevent the oxidation of LDL cholesterol, primarily by scavenging free radicals and other reactive oxygen species in the body's aqueous environment. Moreover, *in vitro* studies have demonstrated that physiological concentrations of ascorbic acid can significantly inhibit LDL oxidation by vascular endothelial cells. The adhesion of leukocytes to the endothelium is a crucial step in initiating atherosclerosis. *In vivo* studies have shown that ascorbic acid can inhibit the interactions between leukocytes and endothelial cells induced by cigarette smoke or OxLDL. These findings suggest that unfermented coconut sap may help reduce the risk of cardiovascular diseases by preventing atherosclerosis and its associated inflammatory processes.

4.5. Nephroprotective Action

In order to overcome the problems of thermal degradation associated with the traditional evaporation process for producing coconut sugar, a low-temperature and pressure process of micronutrient enriched inflorescence sap powder was formulated and registered as COCOZEN™, otherwise referred to as CSP (Coconut Sap Powder) (Svenia et al., 2017). This method not only ensures the preservation of micronutrients in the coconut sugar, but also preserves the sugar-derived bioactive molecules that confer multiple health benefits. CSP also contains flavonoids and nitrates - compounds known to contribute to blood vessel health - and inulin, a prebiotic soluble fiber that helps regulate blood sugar levels. Coconut inflorescence sap has demonstrated significant nephroprotective effects, particularly against gentamicin-induced kidney toxicity. This protective action is attributed to its anti-inflammatory and antioxidant properties, which help to inhibit oxidative stress, lipid peroxidation, and inflammation. By suppressing pro-inflammatory cytokines such as TNF-α and IL-6, CSP helps prevent nephritis and necrosis of kidney tissues (Svenia et al., 2017). These findings open up new possibilities for drug development in managing acute renal diseases, particularly those involving drug-induced kidney damage linked to oxidative stress and lipid peroxidation. Phosphorus, an essential nutrient present in coconut sap, plays a crucial role in maintaining renal health. It aids in the proper elimination of waste from the kidneys through urination and excretion. By increasing the frequency and volume of urination, phosphorus helps the body balance excess salts, uric acid, water, and even fat, as urine typically contains about 4% fat. This process supports the healthy balance of fluids and materials eliminated from the body, thereby helping to keep the entire body healthy and free from toxins (Jose et al., 2017).

Drug-induced nephrotoxicity is a major cause of renal damage. The studies on the nephroprotective and curative effects of CSP in Wistar rats (Jose et al., 2017) highlighted its potential as a natural agent to safeguard and treat kidney related issues. The authors opined that CSP offers opportunities for drug development in the management of acute renal disease, especially drug induced kidney damage with the etiology of oxidative stress and lipid peroxidation. In another study, Jose et al., (2018) observed that supplementation of CSP resulted in significant inhibition of alcoholic hepatic damage with reversal of the biochemical markers and indicated hepatic cell regeneration. The stable powder form has great potential for use as it can overcome the inherent problem of rapid fermentation of the sap to alcohol during storage. The safety and efficacy of CSP (15 g day^{-1} for 90 days) on healthy and non-athletic subjects was investigated (Mammen et al., 2020). The CSP possessed a GI of 52.47 and was safe with an acceptable sensory attribute. There were no side effects or deviations in blood sugar or other hematological and biochemical parameters, except in Hb and lipid profile which showed enhancement. Though CSP contains a high level of carbohydrate (~80% w/w) the beneficial effect of CSP could be due to its rich micronutrient profile.

CSP possessed significant nephroprotective effect on gentamicin induced renal damage by modulating oxidative stress and inflammatory markers. CSP also showed its ability to down regulate and reverse the alterations induced by the oxidized low-density lipoproteins (ox- LDL) and inflammation in human peripheral blood mononuclear cells (hPBMCs) (Ratheesh et al., 2017).

4.6. Hypoglycemia

Individuals with meal-related reactive hypoglycemia often experience excessive insulin secretion after eating, leading to a rapid drop in blood sugar levels. This can result in symptoms like irritability, hunger, weakness, and dizziness. Coconut sap, with its rich nutrient profile—including essential minerals like iron, potassium, magnesium, and zinc, along with vitamins—can help manage this condition. Due to its low GI and glycemic load, coconut sap provides a steady release of energy without causing significant fluctuations in blood glucose levels. By maintaining stable blood glucose and insulin levels, coconut sap can help prevent and manage hypoglycemia and related health issues (Shetty et al., 2023).

4.7. Hepatoprotective Effect

Research conducted at the Indian Institute of Science (IISc), Bengaluru, India suggests that unfermented coconut inflorescence sap may be beneficial for liver health, particularly in cases of alcohol-induced damage. The study highlights acetaldehyde, a toxic byproduct of alcohol metabolism, as a major culprit in liver cell destruction. Coconut sap appears to play a role in eliminating this harmful substance from the liver (Jose et al., 2018). The research suggests the sap may work by influencing specific markers related to inflammation, extracellular matrix breakdown (a process involved in scarring), and oxidative stress (cellular damage caused by free radicals).

Prabha et al., (2019) reported the protective potential of coconut inflorescence sap against cisplatin-induced toxicity in animals. Later, Prabha et al., (2020) found synergistic action of cyclophosphamide and non-fermented coconut inflorescence sap to reduce the solid tumors in mice, possibly through its antioxidant effects. The phenolic-rich acetone extract of coconut inflorescence could be an interesting alternative candidate against acetaminophen-induced hepato-toxicity and associated oxidative stress (Chithra et al., 2020). In another study by Prabha et al., (2021), the non-fermented coconut inflorescence sap ameliorated the cyclophosphamide induced hepatotoxicity attributed through its free radical scavenging activities. Thus, the use of fresh sap may prevent the oxidative stress mediated degenerative ailments.

Treating with paracetamol increases the thiobarbituric acid-reactive substance (TBARS) levels in liver, which enhances lipid peroxidation, leading to tissue damage and failure of antioxidant defense mechanisms. In an animal study, administering 2 ml of coconut inflorescence sap significantly reduced the elevated levels of serum enzymes (Abraham and Raseema 2020). In another study, treatment with coconut inflorescence sap at 25 μg ml^{-1} potentially showed the highest percentage of inhibitory activity against the toxicity induced by paracetamol. There were increased cell viability and also significant decrease in SGOT, SGPT, ALP, LDH, and Glutathione S Transferase (Sharafudeen and Abraham 2024).

Packed with vitamins and minerals, it acts as a nutritional supplement. The amino acid glutamic acid, essential for protein building and nerve function, is abundant in the sap. Phosphorus, another key component, plays a crucial role in digestion by enhancing the body's absorption of vitamins B2 (riboflavin) and B3 (niacin). These vitamins are vital for energy metabolism and neurological health. Beyond vitamin absorption, phosphorus directly alleviates digestive issues like indigestion, constipation, and diarrhea, promoting regular bowel movements. A healthier digestive system also reduces the burden on the kidneys by efficiently eliminating toxins.

4.8. Probiotic Attributes

The probiotic *Lactobacilli* species isolated from coconut inflorescence sap demonstrate notable health benefits, particularly in gut health and immunity. These *Lactobacilli* species are effective in interacting with and colonizing gut epithelial cells. This ability is crucial for maintaining a healthy gut microbiome and ensuring that probiotics can exert their beneficial effects. The probiotic *Lactobacilli* spp present in the sap interacted with the gut epithelium, effectively colonizing gut cells, and such interaction leads to a boost in both adaptive and innate immunity. Among the seven isolates identified from coconut sap, *L. brevis* MYSN 106 was proved to be best, which survived in the low pH and bile conditions in the stomach (Somashekaraiah et al., 2019). Research by Gopal et al., (2021) has shown that these *Lactobacilli* strains enhance both adaptive and innate immunity. This means they not only improve the body's general immune response but also contribute to a more robust and targeted defense against pathogens. Overall, the presence of these probiotic *Lactobacilli* in coconut inflorescence sap contributes significantly to gut health and immune function, making it a valuable addition to diets aimed at improving overall health and immunity.

4.9. Strengthens Bones

Coconut sap is a good source of phosphorus, a mineral essential for strong bones and teeth. Phosphorus works in tandem with calcium to build and maintain bone health.

4.10. Keeps Body Cool

Hydrating and refreshing, coconut sap effectively regulates body temperature due to its electrolyte content. Often recommended for post-surgery recovery, it naturally detoxifies the body without harmful side effects. Its low glycemic index prevents dehydration and provides a cooling effect.

4.11. Ergogenic Potential

Joseph et al., (2021) investigated the ergogenic potential (ability to increase work output or capacity for bodily or mental labour) of CSP produced from coconut inflorescence sap. CSP contains 2 and 5 fold higher levels of vitamins (B and C), 20 and 15 fold electrolytes (Na and K) and around 3, 11, 40 and 16 fold minerals (Ca, P, Fe, and Zn), respectively than coconut water. It also contains 1% nitrate, and was stable and does not undergo fermentation under ambient conditions ($25 \pm 2^{o}C$, relative humidity $60 \pm 5\%$). Administering CSP @3 g day^{-1} for 21 days enhanced the anaerobic sprint power, and running performance in young, apparently healthy participants. Higher blood pressure and arterial stiffness with older age are two important risk factors contributing to cardiovascular disease. Carlini et al., (2023) demonstrated the ability of CSP to lower brachial systolic blood pressure and lessen carotid artery mechanical stiffness in middle-aged and older (≥45 yr) adults.

4.12. Enhances Neurological Function

Coconut sap contains thiamin (vitamin B1) and inositol, both crucial for brain health. Thiamin supports brain energy metabolism, while inositol plays a role in neurotransmitter function. These nutrients in combination contribute to improved neurological function.

In addition to the previously mentioned benefits, coconut sap offers a range of other health advantages. Research indicates that it can be beneficial for treating conditions such as asthma, anemia, bronchial suffocation, tuberculosis, and piles. Coconut sap is rich in inositol, a naturally occurring nutrient that aids in treating eczema and eye abnormalities. Inositol plays a key role in regulating enzyme activity, nerve transmission, and fat transportation within the body. It is also essential for promoting hair growth and can help prevent hair loss and baldness. Furthermore, coconut sap is a natural remedy for kidney stones and urinary issues and is effective in lowering high cholesterol levels. Its antioxidant properties also contribute to its anti-aging effects.

4.13. Anti-Anemic Effect

Fresh and fermented palmyra sap from wild date palms and coconuts are high in iron, indicating that consuming these saps can help alleviate anemia and may serve as an effective remedy for individuals with this condition (Asghar et al., 2020). Additionally, palm honey - a concentrated form of the sap - is also reported to be beneficial for anemic patients due to its high iron content (Hegazi and Abd El-Hady 2009).

Conclusion and Perspectives

The coconut inflorescence sap is a valuable gift from the coconut tree, traditionally revered for its medicinal properties and commonly enjoyed as a refreshing health drink. Notably, it is a non-alcoholic bio-beverage with numerous health benefits and no alcohol content. Globally recognized for its nutritional value and potential for added benefits, this beverage helps cool the body and aids digestion. With an impressively low glycemic index, it does not raise blood sugar levels like conventional fruit juices, making it a safe option for diabetic individuals. Regular consumption of this sap is believed to help prevent diseases such as jaundice and promote overall health. As a natural detoxifying drink, it is praised for its lack of side effects.

Exploring the creation of novel nutritive products from coconut inflorescence sap is a crucial area of research. The high sugar content in this sap offers a valuable substrate for developing products with prebiotic properties, such as fructo-oligosaccharides. Coconut inflorescence sap holds great potential for producing prebiotic-rich products. Additionally, there is a need to investigate the probiotic properties of this sap to develop nutritive products enriched with beneficial probiotic organisms.

When compared to commercially available sugars, coconut sugar stands out as a healthier option. Its fructose and glucose are readily absorbed and converted into energy, whereas sucrose requires initial breakdown into monosaccharides. Additionally, coconut sugar contains higher levels of copper, iron, zinc, and manganese compared to palm and cane sugars. In response to health-conscious consumer preferences, the food industry is increasingly incorporating coconut sugar, sap, and syrup into their products. The process of collecting, transporting, storing, and processing coconut sap requires significant manpower and resources. As a result, the production cost of coconut sugar is higher compared to cane sugar. Future advancements, such as the adoption of AI-based technologies, could help address these cost challenges. Furthermore, research is essential to develop methods for detecting adulteration in coconut sugar, which is crucial for eliminating counterfeit products from the market.

References

Abraham, A., Raseema, S.R. (2020) Hepatoprotective effect of coconut inflorescence sap in paracetamol induced hepatotoxicity. *Sci Technol Dev* 9(2): 80-87, doi:20.18001.STD.2020.V9I2.20.33314.

Aheeshan, B., Maathumai, S., Subajini, M.W., Wijesinghe, A.P.J., Sriwijendran, S. (2022) Antioxidant activity and physicochemical analysis of coconut, kithul and palmyrah palm treacle. *J Sc EUSL* 13(1):23-30, http://doi.org/10.4038/jsc.v13i1.40.

Asghar, M.T., Yusof, Y.A., Noriznan, M., Yaacob, M.E., Ghazali, M.H., Chang, L.S., Manaf, Y.N. (2020). Effect of processing method on vitamin profile, antioxidant properties and total phenolic content of coconut (*Cocos nucifera* L.) sugar syrup. *In J Food Sci Technol* 55(7): 2762–2770. doi: 10.1111/ijfs.14529, 0–2.

Asha, S., Ratheesh, M., Jose, S.P., Krishnakumar, I.M., Sandya, A.N. (2019) *Neera*: A nonalcoholic nutritious beverage from unopened inflorescence of coconut palm. In: Alina MH, Grumezescu G(eds), *Nat. Beverages* Vol. 13 Sci. Beverage, Woodhead Publishing, pp 339–360, https://doi.org/10.1016/B978-0-12-816689-5.00012-2.

Beegum, S.P.P., Jwala, P.N., Manikantan, M.R., Pandiselvam, R., Shil, S., Neenu, S., Hebbar K.B. (2022). Effect of coconut milk, tender coconut and coconut sugar on the physicochemical and sensory attributes in ice cream. *J Food Sci Technol* 59(7): 2605-2616, doi:10.1007/s13197-021-05279-y.

Bhagya, D., Soumya, G. (2016). Effects of coconut neera (*Cocos nucifera* L.) on blood pressure among hypertensive adult women. *Int J Applied and Pure Sci Agric* 2(9):1-7.doi:10.26717/BJSTR.2018.11.002114.

Borse, B.B., Rao, L.J.M., Ramalakshmi, K., Raghavan, B. (2007).Chemical composition of volatiles from coconut sap (Neera) and effect of processing. *Food Chem* 101(3): 877–880.doi:10.1016/j.foodchem.2006.02.026.

Carlini, N.A., Romanowski, S., Rabalais, E.N., Kistler, B.M., Campbell, M.S., Krishnakumar, I.M., Harber, M.P., Fleenor, B.S. (2023) Coconut sugar derived from coconut inflorescence sap lowers systolic blood pressure and arterial stiffness in middle-aged and older adults: a pilot study. *J Appl Physiol* 134(3):508–514.doi: 10.1152/japplphysiol.00394.2022.

Chen, W., Zhu, Q., Xia, Q., Cao, W., Zhao, S., Liu, J. (2011). Reactive oxygen species scavenging activity and DNA protecting effect of fresh and naturally fermented coconut sap. *J Food Biochem* 35: 1381–1388. doi:10.1111/j.1745-4514.2010.00506.x.

Chithra, M.A., Ijinu, T.P., Kharkwal, H., Sharma, R.K., Janardhanan, K.K., Pushpangadan, P., George, V. (2020) *Cocos nucifera* L. inflorescence extract: An effective hepatoprotective agent. *Indian J Traditional Knowledge* 19(1):128-136.

Devi S.N., HariPrasad T., Ramesh K., Ramchander M. (2015). Antioxidant properties of coconut sap and its sugars. *Int J Pharm Tech Res* 8: 160–162.

Femi, B.M.R., Malhija N., Sherin Mary T., Kumaran T. (2020) Coconut (*Cocos nucifera)* inflorescence sap nutritional characteristics and health effects. *J. Basic Appl. Res Biomed* 6(1): 50-52.

Ghosh, D.K., Bandyopadhyay A., Baidyanath D., Kaushik G., Hebbar K.B., Chowdappa P. (2016). Method of the preparation of fresh coconut inflorescence sap (*Kalparasa*) based sweets. Patent Application No. 201631013187.

Gopal, M., Shil, S., Gupta, A., Hebbar, K.B., Arivalagan, M. (2021). Metagenomic investigation uncovers presence of probiotic-type microbiome in *Kalparasa* (fresh unfermented coconut inflorescence sap). *Frontiers Microbiol*, 12. doi:10.3389/fmicb.2021.662783.

Hali, R. (2013). Bio beverage – Coco Neera. *Indian Coconut J.* LVI 17–19.

Haryanti, P., Karseno, K., Setyawati, R. (2018) Natural Preservatives Application Form Nira Coconut Powder Made Against Green Betel Physical and Chemical Properties Coconut Sugar. *Pembangunan Pedesaan* 1 (1):103-109.

Haryanti, P., Supriyadi, S., Marseno, D.W. and Santoso, U. (2021) The effect of arginine addition on chemical and antioxidant properties of coconut sap during heating. *OP Conf. Series: Earth and Environmental Science* 653. doi:10.1088/1755-1315/653/1/012143.

Haryanti, P., Supriyadi, S., Marseno, D.W. and Santoso, U. (2018) Effects of different weather conditions and addition of mangosteen peel powder on chemical properties and antioxidant activity of coconut sap. *Agritech* 38 (3):295-303, https://doi.org/10.22146/agritech.29844.

Hebbar, K.B., Arivalagan, M., Manikantan, M.R., Mathew, A.C., Chowdappa, P. (2015a). *Kalparasa collection and value addition*. Technical bulletin No. 92, ICAR-CPCRI, Kasaragod, Kerala.

Hebbar, K.B., Arivalagan, M., Manikantan, M.R., Mathew, A.C., Thamban, C., Thomas, G.V., Chowdappa, P. (2015b). Coconut inflorescence sap and its value addition as sugar—collection techniques, yield, properties and market perspective. *Current Sci* 109(8): 1411–1417, doi:10.18520/v109/i8/1411-1417.

Hebbar, K.B., Arivalagan, M., Pavithra, K.C., Roy, T.K., Gopal, M., Shivashankara, K.S., Chowdappa, P. (2020). Nutritional profiling of coconut (*Cocos nucifera* L.) inflorescence sap collected using novel coco-sap chiller method and its value added products. *J Food Measurement and Characterization* 14 (5):2703–2712. doi:10.1007/s11694-020-00516-y.

Hebbar, K.B., Augustine. (2021). 'Cocosap chiller' *Patent* No373309.

Hebbar, K.B., Pandiselvam, R., Manikantan, M.R., Arivalagan, M., Beegum, S.P.P., Chowdappa, P. (2018). Palm sap—quality profiles, fermentation chemistry, and preservation methods. *Sugar Tech* 20(2): 621–634. doi:10.1007/s12355-018-0597-z.

Hebbar, K.B., Ramesh, S.V., Ghosh, D.K., Beegum, S.P.P., Pandiselvam, R., Manikantan, M.R., Mathew, A.C. (2022). Coconut sugar-a potential storehouse of nutritive metabolites, novel bio-products and prospects. *Sugar Tech* 24(3):841-856. doi:10.1007/s12355-021-01101-3.

Hegazi, A.G., Abd El-Hady, F.K. (2009). Influence of honey on the suppression of human low density lipoprotein (LDL) peroxidation (in vitro). *Evidence-based Complementary and Alternative Med* 6(1):113–121. doi:10.1093/ecam/nem071.

Jose, S.P., Asha, S., Krishnakumar, I.M., Ratheesh, M., Santhosh, S., Sandya, S., Kumar G., Pramod, C. (2017). Nephro-protective effect of a novel formulation of unopened coconut inflorescence sap powder on gentamicin induced renal damage by modulating oxidative stress and inflammatory markers. *Biomed Pharmacother* 85: 128-135. doi: 10.1016/j.biopha.2016.11.117.

Jose, S.P., Mohanan, R., Sandya, S., Asha, S., Krishnakumar, I.M. (2018). A novel powder formulation of coconut inflorescence sap inhibits alcoholic liver damage by modulating inflammatory markers, extracellular matrix metalloproteinase, and oxidative stress. *J Food Biochem* 42(5), p.e12543.doi:10.1111/jfbc.12543.

Joseph, A., Jose S.P., Kalyan B.T., Mammen R.R., Krishnakumar I.M., Fleenor R.S., Mohan R. (2021) Coconut inflorescence sap enhances exercise performance and plasma antioxidant status in young active men. *NFS Journal* 23: 37–43, https://doi.org/10.1016/j.nfs.2021.03.002.

Mammen, R., John, F., Krishnakumar, I.M., Maliakel, B., Mohanan, R., Fleenor, B. (2020). Safety and influence of a novel powder form of coconut inflorescence sap on glycemic index and lipid profile. *Bioact Carbohydr Diet Fibre* 23, 100217, https://doi.org/10.1016/j.bcdf.2020.100217, 7 pages.

Martins, S.I.F.S., Jongen, W.M.F., van Boekel, M.A.J.S. (2001). A review of Maillard reaction in food and implications to kinetic modeling. *Trends Food Sci Technol* 11(9-10): 364–373doi:10.1016/S0924-2244(01)00022-X.

Mishra, B. (2016). Neera: The coconut sap: a review. *Int J Food Sci Nutr* 1:35–38.

Panonnummal, R., Gopinath, D. Thankappan Presanna, A., Viswanad, V., Mangalathillam, S. (2021). Non-alcoholic palm nectar from *Cocos nucifera* as a promising nutraceutical preparation. *J. Food Biochem* 46, e13900, doi: 10.1111/jfbc.13900.

Prabha, S., Meera, N., Mathew, S.E., Indu, M.S., Ramavarma, S.K., Kuzhuvelil, B.T., Devassy, B.T. (2021) *Neera*, a non-fermented traditional drink from coconut spadix restores the redox status in sodium fluoride intoxicated mice. *J Complementary and Int Med* 18(3)doi: 10.1515/jcim-2020-0074.

Prabha, S., Nair, M., Kuzhivelil, B.T., Babu, T.D. (2017) Analysis of antioxidant and anti-inflammatory activity of coconut inflorescence sap (CIS). Presented in the 36th Annual Conference of Indian Association for Cancer Research on: The 21st Century – War on Cancer, February 9-11, 2017, Amala Cancer Research Centre, Thrissur, Kerala, India.

Prabha, S., Shaji, E.M., Kuzhivelil, B.T., Babu, T.D. (2020) *Neera*, non-fermented coconut inflorescence sap on alleviated cyclophosphamide induced systemic toxicity in mice. 2nd National Conference on Current Trends in Biological Sciences At: St Xavier's College, Palayamkottai, Tamil Nadu, February 6 -7, 2020.

Prabha, S., Shaji, E.M., Suseela, I.M., Kuzhivelil, B.T., Babu, T.D. (2019) Protective effect of Neera, non- alcoholic coconut inflorescence sap on cisplatin induced nephrotoxicity in mice. Presented in the International Seminar on 'Modern Trends in Biological Sciences' On 5th February 2019 Organized by Department of Biochemistry & IQAC Sree Krishna College Guruvayur- Thrissur 680102.

Ratheesh, M, Svenia, JP, Asha, S, Sandya, S, Girishkumar, B, Krishnakumar, IM (2017). Anti-inflammatory effect of a novel formulation of coconut inflorescence sap against ox-LDL induced inflammatory responses in human peripheral blood mononuclear cells by modulating TLR-NF-κB signaling pathway. *Toxicol Mech Methods* 27(8):1-22. doi: 10.1080/15376516.2017.1344339.

Raveendran, S.R., Thankappan, R. (2012). Protective and curative effects of *Cocos nucifera* inflorescence on alloxan-induced pancreatic cytotoxicity in rats. *Indian J Pharm* 44: 555–559.doi: 10.4103/0253-7613.100368.

Renjith, R.S., Chikku, A.M., Rajamohan, T. (2013) Cytoprotective, antihyperglycemic and phytochemical properties of *Cocos nucifera* (L.) inflorescence. *Asian Pacific J Tropical Med* 6(10):804-810, doi: 10.1016/S1995-7645(13)60142-X.

Renjith, R.S., Rajamohan, T (2012a) Protective and curative effects of *Cocos nucifera* inflorescence on alloxan-induced pancreatic cytotoxicity in rats. *Indian J Pharm* 44(5):555–559. https://doi. org/10.4103/0253-7613.100368.

Renjith, R.S., Rajamohan, T. (2012b) Young inflorescence of *Cocos nucifera* contributes to improvement of glucose homeostasis and antioxidant status in diabetic rats. *Int J Diabetes Dev Countries* 32(4):193–198, doi:10.1007/s13410-012-0088-9.

Sharafudeen, R.R., Abraham, A., Soman, R.U. (2022) Phytochemical profiling and antioxidant potential of coconut inflorescence sap. *Int J Pharm Investigation* 12(3) : 255-259, doi: 10.5530/ijpi.2022.3.46.

Sharafudeen, R.R., Abraham, A. (2024) Hepatoprotective potential of coconut inflorescence sap against paracetamol induced toxicity in hep G2 cell lines. *Food Chem Toxicol* 193, 114946, https://doi.org/10.1016/j.fct.2024.114946.

Shetty. S.S., Ramesh, S.V., Arivalagan, M., Roopashree, P.G., Manikantan, M.R., Hebbar, K.B., Kumari, S. (2023) Coconut (*Cocos nucifera* L.) inflorescence sap-derived sugar restores the glucose and lipid homeostasis in streptozotocin-induced diabetic Wistar rat model. *Biomed* 43(3):859–866 doi:10.51248/.v43i3.1453.

Silpa, P., Meera, N., Shaji, E.M., Indu, M.S., Smitha, K.R., Kuzhivelil, B.T., Babu, T.D. (2021). *Neera*, a non-fermented traditional drink from coconut spadix restores the redox status in sodium fluoride intoxicated mice. *J Complementary and Integrative Med* 18(3): 499-505.doi: 10.1515/jcim-2020-0074.

Somashekaraiah, R., Shruthi, B., Deepthi, B.V., Sreenivasa, M.Y. (2019) Probiotic properties of lactic acid bacteria isolated from neera: a naturally fermenting coconut palm nectar. *Frontiers in Microbiol* 10: 1–11, doi:10.3389/fmicb.2019.01382.

Sulistyo, S. B. and Haryanti, P. (2020) Regression analysis for determination of antioxidant activity of coconut sap under various heating temperature and concentration of lysine addition. *Food Res 4*(4):976–981, doi: 10.26656/fr.2017.4(4).410.

Svenia, J.P., Asha, S., Krishnakumar, I.M., Ratheesh, M., Savitha, S., Sandya, S., Girish, K.B., Pramod, C. (2017). Nephro-protective effect of a novel formulation of unopened coconut inflorescence sap powder on gentamicin induced renal damage by modulating oxidative stress and inflammatory markers. *Biomed Pharmacother* 85: 128–135.doi: 10.1016/j.biopha.2016.11.117.

Thampan, P.K. (2013). Production of neera and coconut sugar deserves encouragement. *Indian Coconut J* 11–14.

Valko, M., Rhodes, C.J., Monocol, J., Izakovic, M., Mazur, M. (2006). Free radicals, metals and antioxidants in oxidative stress-induced cancer. *Chem Biol Interact* 160: 1–40 doi:10.1016/j.cbi.2005.12.009.

Wrage, J., Burmester, S., Kuballa, J., Rohn, S. (2019). Coconut sugar: Production process, chemical characterization, and sensory properties. *LWT—Food Sci Technol* 112:108227, doi: 10.1016/j.lwt.2019.05.125.

Chapter 10

Coconut Palm Sap: Consumer Preference, Income Prospects and Employment Generation

S. Jayasekhar*
ICAR-Central Plantation Crops Research Institute, Kasaragod, India

Abstract

This chapter illustrates the global trade dynamics, socio-economic importance, and employment generation potentials of coconut palm sap and palm sugar. Analyzing the newfound opportunities of coconut palm sap and sugar is crucial for the coconut-growing countries seeking to leverage emerging possibilities in the international market. Regions such as Europe and America have significant market potential due to an increasing population of health-conscious consumers seeking healthy and organic food choices. The global value chain of coconut palm sap originates from the regions of South Asia and ends with specialized market segments such as fair-trade, organic certification, and nutraceuticals in the EU and US. The global value chain of coconut sugar is currently characterized as a buyer-driven chain, primarily because a few dominant players in the export destinations control the entire value chain. In the production node of the value chain, major hurdles are scarcity of palm climbers, inadequate infrastructure, lack of processing facilities, and market-related constraints such as product uniformity across the space. The market strategy plays a crucial role in the success of coconut palm sap industry. The product must compete with other packed fruit juices and sodas, and due to its superior nutritional value

* Corresponding Author's Email: jaycpcri@gmail.com

In: Coconut Palm Sap
Editors: K. B. Hebbar, V. Krishnakumar and S. V. Ramesh
ISBN: 979-8-89530-640-6

compared to other substitutes, it must be strategically positioned to attract consumers. Coconut sugar manufacturers can position their product lines in premium categories to take advantage of the growing demand for organic and fair-trade products worldwide. To get market access in the higher-end sectors, it is important to promote coconut sugar as a natural product, distinguishing it from traditional white sugar and synthetic sweeteners.

Keywords: coconut palm sap, coconut sugar, trade dynamics, value chain, fair-trade

1. Introduction

Coconut growers are currently more exposed to economic risks and uncertainties owing to the high degree of price fluctuations. To maintain the growth momentum of coconut exports, it is critical to innovate and explore the untapped segments (Jayasekhar and Thomas 2024). The coconut wood, coconut gratings, tender coconut water, coconut sugar, coconut jaggery, coconut vinegar, coconut sap etc. are some of the coconut value-added products that have high demand in niche markets, and diversification can help to suitably leverage these markets. Coconut sap is the latest addition to the array of value-added products that hold a significant potential to revitalize coconut farmers' prospects. The phloem sap is abundant in sugars, amino acids, minerals, vitamins, and antioxidants. Coconut palm sap serves as the primary ingredient for several value-added products such as palm syrup, palm jaggery, and palm sugar, which are in significant demand both locally and globally due to their substantial nutritional benefits. The product's most notable feature is its low glycemic index (GI), which indicates the degree of sugar absorption into the bloodstream. The worldwide demand for low-GI sugar is increasing, and hence coconut sap and coconut palm sugar as natural sweetener alternatives can address this need.

The coconut palm sugar is the widely known commercial product derived from the coconut inflorescence sap. The most significant factor contributing to lifestyle-related illnesses is the excessive consumption of sweets and artificial sweeteners, both of which pose considerable health risks. Coconut sugar, with a lower calorific content than conventional sugar, has several applications in food industry. The appeal of coconut sugar is driven by its elevated levels of essential vitamins and minerals, including magnesium, potassium, and zinc. The worldwide coconut sugar market is anticipated to grow from USD 3635 million in 2023 to USD 4936 million by 2033. The coconut sugar market is projected to experience a 3.6% compound annual growth rate (CAGR) from 2023 to 2033 (Choudhury 2023).

In case of coconut palm sugar, the current market has several regional and local participants. Approximately 75% of the market is dominated by domestic entities. Regions such as Europe and America have significant market potential due to an increase in health-conscious consumers seeking healthy and organic food choices. However, it is striking that, owing to the elevated price of palm sugar, multinational corporations own a minimal market share of less than 10% relative to local rivals (Choudhury 2023; Kearney 2010).

Analyzing the worldwide competitiveness, performance, and trade dynamics of coconut palm sap and sugar is crucial for any coconut-producing nation seeking to leverage emerging possibilities in the international market. In this context, there is an imminent need for an extensive investigation on the intricacies of worldwide coconut palm sap and sugar spectrum. This chapter aims to depict the global landscape of coconut palm sap sector by detailing trade, global value chain, market access, and the possibility for rural empowerment within the sector.

2. World Scenario of Coconut

India, the Philippines, and Indonesia together contributed 73.6% of global coconut production, which totalled 67,019 million nuts (Table 1). India remains the world's leading producer of coconuts, followed closely by the Philippines and Indonesia. The global coconut area stood at 12.2 million ha, with the Philippines holding highest area of 3.6 million ha, followed by Indonesia at 3.34 million ha and India at 2.11 million ha. Approximately 75% of worldwide coconut acreage and production in 2022 was attributed to these top three coconut growing countries. In 2022, global coconut production declined by over 1.5% compared to 2021, while the area under coconut cultivation decreased by approximately 1.3% during the same period. Among the top three producers, India recorded the highest yield at 9,122 nuts per hectare, followed by Indonesia and the Philippines with yields of 4,274 and 4,156 nuts per hectare, respectively. The global average yield stood at 5,429 nuts per hectare. Over the past five years, India's share of global coconut production has decreased from 34.3% in 2018 to 28.8% in 2022. In contrast, the Philippines' share has risen from 21.2% to 22.3%, while Indonesia's share has improved from 20.5% to 22.5% over the same period.

Table 2 displays the country-wise export shares of major coconut value-added products. It is surprising that the Philippines and Indonesia together accounted for about 78% of global coconut oil exports. Similarly, in case of desiccated coconuts, these two countries hold the major export share. Indonesia and Papua New Guinea account for 60% export share in the globally traded copra. Vietnam and India lead the way in coconut sugar exports, collectively contributing 98% of global export share. Vietnam is the major contributor of coconut milk exports, followed by Indonesia and Sri Lanka. It is observed that global competition in each coconut product line is skewed, which may detrimentally affect the price scenario of coconut value-added products in the event of a supply deficit from major exporting country. For example, though India is the world's leading coconut producer, it lags behind other coconut-growing countries in terms of export share of value-added products. In the international trade landscape, each country should identify its respective comparative advantages in coconut product exports and develop strategies to expand market access.

Table 1. World coconut area and production

Country	Area ('000 ha)	Area (% share)	Production (million nuts)	Production (% share)	Yield (nutsha^{-1})
Indonesia	3342	27.3	15048	22.5	4274
India	2154	17.6	19310	28.8	9122
Philippines	3604	29.5	14931	22.3	4156
Others	3129	25.6	17730	26.5	---
World	12229	100	67019	100	5429

Source: ICC, 2024.

Table 2. Global scenario of coconut products exports (percentage share)

Country/Item	Copra	Coconut Oil	Desiccated Coconut	Coconut Sugar	Coconut Milk/Cream
Philippines	---	49.73	31.39	0.07	5.51
Indonesia	32.78	28.11	22.14	1.73	29.62
Sri Lanka	---	---	8.78	---	17.51
Vietnam	---	---	4.84	44.54	---
Malaysia	---	5.36	2.59	---	---
India	9.89	0.82	1.49	53.55	---
Papua New Guinea	27.11	1.20	---	---	---
Thailand				0.11	47.35

Source: ICC, 2024 and UNCOMTRADE, 2024.

3. Economic Importance of Coconut Palm Sap and Sugar

The economic importance of coconut palm sap and palm sugar, along with their employment potential and possible rural empowerment has been illustrated in this section based on an array of literature reviews. Annual coconut sugar production in Indonesia is approximately 300,000 tonnes (Novarianto et al., 2021). Annually, the demand for coconut sugar in Indonesia, particularly as a primary ingredient for manufacturing of sweet soy sauce, showed a 10% growth. In the recent months, the supply of raw material (coconut sugar) for sweet soy product has declined. The scarcity of coconut climbers collecting coconut sap is the major reason for deficit supply of coconut sap. It could be attributed to several reasons, one of which being the lack of enthusiasm among young tappers in ascending tall coconut palms. In order to address this issue, it is necessary to develop dwarf coconuts, characterised by their shorter trunk and early fruiting, suitable for sap tapping.

Another classic example of coconut sap farmers from the Karangsari village of Adipala district, Indonesia proved that the effective empowerment process strengthens the existing local farmer organizations and steer them towards the path of sustainable income generation (Trianingrum et al., 2023). The farmers of this village have entered into the global value chain of coconut sugar by processing and marketing the best quality coconut sugar to the high value European destinations. The study conducted in the West Sei Kepayang district provides further insights into the economic feasibility of the coconut sugar industry at the village level. The project achieved an adequate net present value over a period of 20 years. The internal rate of return (12.31%) exceeded the prevailing credit interest rates in the market. The benefit-cost ratio was determined to be 2.60, with a payback period of three years (Matondang et al., 2017).

Another study from the Philippines serves as a testament to the commercial success of coconut palm sap and sugar production at the smallholder level. It was irrefutably evident that coconut producers have access to promising opportunities to improve their farm income. The profitability analysis of sugar production as an income-boosting enterprise demonstrates that a farmer can earn roughly USD 11,000 annually from 100 coconut trees on a one-hectare coconut farm. Sensitivity analysis indicates that the enterprise remains profitable even with a price reduction of 10–20% (Manohar et al., 2009). Despite a 16% decrease in fluid supply from an average assumption of 3 L $tree^{-1}$ day^{-1} and a 10% decrease in the cost of sugar, net income would decrease by only 50%. This economic analysis indicates that the coconut sugar enterprise is a commercially viable farm-level technology.

The employability, profitability and the untapped potential of coconut palm sap was highlighted in Indian scenario based on a study conducted in West Bengal, India. A farmer with 100 coconut palms can generate eight heads of employment annually with an average net return of Rs.9.24 lakhs (Ghosh et al., 2018). It is also noteworthy that in West Bengal, there is a growing demand for palm sap-based milk sweets, palm sugar, and palm jaggery, which are being manufactured and marketed as value-added products.

The profitability and employment generation potential of coconut sugar agro-industry has been documented by a study at the Sukanagara village, Lakbok, Indonesia (Tito Hardianto 2020). The coconut sugar sector significantly contributes to employment, especially in areas where it is a sole agricultural venture. The potential for establishing coconut sugar agro-business in Sukanagara village is encouraging, as the industry achieved a profit of 29% over the invested capital in a single production cycle. Furthermore, employment in coconut sugar sector constitutes 2.47% of the workforce, significantly impacting local job opportunities.

Consumer perception research done in India found that respondents' understanding of coconut palm sap is below average, but their satisfaction from the product is above average. It demonstrates that the awareness on health benefits and non-alcoholic properties of sap drink will increase the consumer satisfaction with the beverage. Because there is significant misunderstanding and misinformation about palm sap among consumers, raising awareness of its advantages and dispelling misconceptions is crucial. As a result, it is recommended that coconut palm sap marketers focus on increasing the quality and brand image of the drink, while also prioritizing efforts on sales promotion efforts and awareness programs (Sreeresh 2019).

An attempt to outline the reasons for setbacks in coconut palm sap sector and also to provide a potent framework for revamping the sector was performed in Kerala, India (Thamban et al., 2020). The sap from coconut inflorescences has been advocated as a promising value-added non-alcoholic beverage with numerous health benefits in the state. The collaborative endeavour of stakeholder agencies ensured exemption of the goods from the production regulations imposed by the excise statute. Consequently, the coconut producer federations (CPFs) in the state received licenses for manufacturing and distribution of palm sap. Despite the optimistic forecasts for the industry, most federations have ceased the palm sap venture, which they initially pursued with much enthusiasm. The palm sap value chain is currently in an emerging stage, and absence of institutional backing from the government has adversely impacted the trust of the stakeholders' engaged in it. The deficiency of technical expertise and absence of marketing abilities were clearly apparent in the sector. The scarcity of palm sap technicians (tappers) and elevated wage rates have raised concerns over profitability and reliability of continuing product supply. The study unequivocally emphasises the necessity for a reconfigured value chain of coconut palm sap, with distinct functions allocated to stakeholders via the establishment of functional links.

4. Global Trade and Value Chain of Coconut Palm Sap and Sugar

People across the world are becoming more health conscious, leading to a significant demand for natural beverages with high nutritive values and rich energy content. In this regard, coconut palm sap is quite important economically, particularly for people living in rural and suburban areas. The main producers of coconut palm sap are the Philippines, India, Vietnam, Thailand, and Sri Lanka. Statistics of coconut palm sap exports reveal that South Korea, the United States, Canada, Japan, Norway, France, Australia, and the United Arab Emirates are the main export destinations (Sarma et al., 2022). The World Health Organisation (WHO 2023) estimates that 422 million people worldwide suffer from diabetes, and in order to maintain low blood glucose, they need to use an alternate sweetener, ideally one that is natural. Because palm saps are classified as low-GI foods, their market value has the potential to grow significantly, both for the sap itself and for value-added products derived from it. For instance, this nutrient-rich sap contributes USD 105 million annually to India's GDP, making a substantial impact on the nation's economy (Jnanadevan, 2015). A farmer with 100 palm trees can generate an average net return of USD 13,000 per year while also providing employment opportunities within the local community (Ghosh et al., 2018). In India, value addition and tapping of coconut palm sap primarily cater to the domestic market, whereas countries such as Indonesia, the Philippines, Malaysia, Sri Lanka, Vietnam, and Thailand serve the global market. However, coconut sugar export data presents a different perspective altogether (Table 3).

Table 3. World exports of coconut sugar, 2017 - 2022 (tonnes)

Country/Year	2017	2018	2019	2020	2021	2022
Vietnam	611	1341	2120	60063	20281	33416
Indonesia	1067	1579	1495	2905	2022	1296
India	6460	10606	12306	11455	18806	40178
Philippines	847	1611	857	983	976	49
Total	8985	15137	16778	75406	21804	75025

Source: ICC-Coconut statistical year book-2022.

Since global databases on agricultural production and trade flows do not currently track coconut sugar as a distinct commodity from other sweeteners, obtaining accurate estimates of market size and trade volumes is challenging. However, according to the International Coconut Community, total world exports in 2022 were estimated at 75,025 tonnes, with India leading at 40,178 tonnes, followed by Vietnam at 33,416 tonnes. Trade instability from 2017 to 2022, highlights that the global market for coconut sugar is still in its early stages and requires strategic efforts to achieve stability.

In developing nations across South Asia and Africa, rural industrialization is crucial for generating new employment opportunities. The processing of palm sap could be a key driver for industrialization in rural areas. However, marketing strategies must align with region-specific sap availability. To maximize economic value and support the palm sap-based food industry, sustainable technologies should be adopted in alignment with the regional demographic factors. The World Health Organisation recommends that sugar intake should not exceed 10% of total caloric consumption. Only a few nations achieve this required standard. However, the consumption of sugar in Europe is anticipated to decline in the future. In 2021, the European soft drinks sector declared a commitment to decrease added sugars in soft drinks by an additional 10% from 2019 to 2025. Notably, this commitment excludes coconut sugar. However, it indicates a trend towards reducing sugar intake as consumer knowledge in Europe rises. Meanwhile the coconut sugar is attaining increasingly popular in the European market. At present, the coconut sugar market in Europe is around 2,000 tonnes, and the organic coconut sugar market is projected to expand at a compound yearly growth rate of 6% from 2020 to 2025 (CBI 2022). The above premises offer huge market access potential for coconut sugar producing countries in the future.

The major global value chain of coconut sugar starts from farmer in the coconut growing country. Different countries have different framework of sap collection. For instance, in India, the producer collectives are the upstream actors, while in Indonesia, it is an individual farmer. However, the succeeding node in the value chain is local processing and subsequently it passes through the exporter. While considering the high value markets such as EU and US, the chain continues along the destination market traders such as importers, brokers and agents. Further the value chain continues through sugar packers and food manufacturers. Towards the end, the value chain ends mostly in organic segments or in fair trade segments, and at sometimes ends with the niche segment of diabetes. Since coconut sugar in high value markets focus on speciality segments, it is imperative for the upstream sap collection hubs in the coconut growing regions to comply with the food safety requirements with utmost caution. The global value chain of coconut sugar, as of now is characterized as a buyer driven chain mainly because a few dominant players in the export destinations are controlling the entire value chain, and also it is necessary to comply with the certifications and standards of fair trade and organic segments in the end markets.

Conclusions and The Way Forward

Arguably the most pressing issue concerning the production of coconut palm sap on a commercial scale is the scarcity of skilled tappers, especially in the South Asian countries, where the palms are predominantly tall in nature. Furthermore, lack of adequate technical support along various nodes of coconut palm sap chain and inadequate infrastructural facilities, that include processing facilities, required to be addressed. The success of coconut palm sap heavily relies on the market strategy, wherein the product has to compete with other packed fruit juices and sodas. Palm sap is nutritionally superior to other substitutes and hence it has to be appropriately positioned to evoke special attraction. The lack of product consistency is a major weakness, as product uniformity across the space is an important criterion for market penetration. It is necessary that the farmer producer organizations engaged in coconut palm sap sector need continuous handholding support to help them deal with the technical and marketing challenges in enterprise development. There should also be mechanisms to respond to policy and institutional bottlenecks that can constrain the scaling up of new farmer enterprises.

A well-structured, targeted entrepreneurial development program focussing on various aspects of the palm sap value chain would assist the trade aspirants in overcoming the inertia of entrepreneurship during the initial phase. The functional connections among major stakeholders in coconut palm sap innovation system must be reinforced with clearly defined responsibilities to enhance profitability and sustainability. Enhancing value share in the production node through potential chain integration or upgradation is critically important.

Coconut sugar, being a healthier alternative to regular sugar, holds significant promise in dietary products. Coconut sugar producers from developing countries may capitalise on the high-value segments of EU and US markets by marketing their product as a premium and healthier alternative to traditional sugar. Coconut sugar producers may leverage the demand of organic and fair-trade products in these markets to position their product lines in premium segments. To get market access in the higher-end sectors, it is important to promote coconut sugar as a natural product, distinguishing it from traditional white sugar and synthetic sweeteners. It is critical to highlight the health benefits of coconut sugar when compared to traditional sugars. It is also crucial to educate the customers about its use as a healthy alternative, as there exists a huge knowledge gap in this arena.

References

CBI (2022). *The European market potential for coconut sugar. Market study, Centre for the Promotion of Imports from developing countries.* https://www.cbi.eu/market-information/natural-food-additives/coconut-sugar/market-potential.

Choudhury, N.R. (2023). Coconut Sugar Market Outlook (2023 to 2033). REP-GB-17817. *Future Market Insights Inc,* USA, 350p. https://www.futuremarketinsights.com/reports/coconut-sugar-market.

Ghosh, D.K., Bandyopadhyay, A., Das, S., Hebbar, K.B., Biswas, B. (2018). Coconut sap (Neera)-Untapped opportunity of spinoff gains in West Bengal, India. *Int J Current Microbiol Applied Sci.* 7(9): 1883-1897.

ICC (2024). Coconut Statistical Year Book-2022, *International Coconut Community,* Jakarta, Indonesia, ISSN 0855-235X: 200p.

Jayasekhar, S., Thomas V. (2024). Policy issues and trade potential of coconut: A global outlook. In: Ramesh S, Praveen S (eds) *Coconut-Based Nutrition and Nutraceutical Perspectives*. Springer, Singapore. https://doi.org/10.1007/978-981-97-3976-9_12.

Jnanadevan, R. (2015). Neera tapping for making coconut the most profitable. *Cocoinfo Int* 22(1): 20–22.

Kearney, J. (2010). Food consumption trends and drivers. Philosophical transactions of the *Royal Society B: Biol Sci* 365: 2793-2807. https://doi.org/10.1098/rstb.2010.0149.

Manohar, E.C., Kindipan, N.M.L., Sancha, L.V. (2009). Coconut sap sugar production: from farm to market and from wealth to health. *Philippine Entomologists* 32 (Supplement no. 1): p. 20.

Matondang, K.A., Suman, A., Sasongko., Mahardika P.A.S. (2017). Local economic development-based on farming manufacture of coconut sugar in West Sei Kepayang district. *Int J Social and Local Economic Governance* 3 (1): 49-55.

Novarianto, H., Tulalo, M.A., Mawardi, S. (2021). Coconut sugar production of Dwarf coconut varieties. *IOP Conf. Series: Earth and Environmental Science* 800,012024, doi:10.1088/1755-1315/800/1/012024.

Sarma, C.A.B., Mummaleti, G.A., Sivanandham, V.C., Kalakandan, S.A., Rawson, A.D., Anandharaj, A.D. (2022). Anthology of palm sap: The global status, nutritional composition, health benefits and value added products. *Trends in Food Sci Technol* 119: 530-549.

Sreeresh, D. (2019). The consumer centric study of neera; A coconut treasure house Of untapped potentials. *Online Int Interdisciplinary Res J* 9(1): 102-107.

Thamban, C., Jayasekhar, S., Chandran, K.P., Rajesh, M.K. (2020). Sustainability of farmer producer organisations - The case of producer organisations involved in the production and marketing of 'neera' in the coconut sector. *J Plantn Crops* 48(2):150-159.

Tito Hardianto (2020). Profitabilitas dan PeluangPengembangan Agroindustri Gula Kelapa dalamSistemAgribisnis Kelapa (*Cocos nucifera* L.). Agritekh (JurnalAgribisnis dan Teknologi Pangan). *Mei* 2020. 1(1): Halaman 46-58.

Trianingrum, S., Arifin, M., Primatika, Anggoma F.F., Mubarok, A.M. (2023). Prosperity with sugar: A strategy for sustainable livelihoods for coconut sap farmers in the Pendekar CSR program. *Int J Sci Society* 5 (2): 259-274.

WHO (2023). Report of the fifth meeting of the WHO Technical Advisory Group on Diabetes: hybrid meeting, 7–8 June 2023. *Geneva: World Health Organization*; 2023. Licence: CC BY-NC-SA 3.0 IGO. ISBN 978-92-4-008482-7, 26p.

Chapter 11

Coconut Palm Sap – The Way Forward

K. B. Hebbar*
V. Krishnakumar
and S. V. Ramesh
ICAR-Central Plantation Crops Research Institute, Kasaragod, Kerala, India

Abstract

Tapping coconut inflorescence sap and its sale as fresh beverage or converting into multiple value-added food products is highly profitable compared to the sale of coconuts. Nonetheless, need for efficient climbing or tapping solutions and concerns with respect to its preservation and improvement of shelf life remain. There is lack of consumer awareness about its proven health benefits, and hence, more clinical studies endorsing the plethora of potential health benefits and popularization of the sap are warranted. Adulteration of sap derived sugar is a major area of concern warranting R & D intervention to devise suitable and simple detection technologies. In this context, the final chapter of this anthology discusses the way forward for coconut inflorescence sap.

Keywords: sustainability, ethical sourcing, rural economy, value addition

* Corresponding Author's Email: hebbar.kb@icar.gov.in

In: Coconut Palm Sap
Editors: K. B. Hebbar, V. Krishnakumar and S. V. Ramesh
ISBN: 979-8-89530-640-6

1. Introduction

Coconut inflorescence sap or '*neera,*' is a sweet, nutrient-rich sap extracted from the unopened flower of coconut palm. Traditionally consumed fresh or fermented into products like toddy, vinegar or processed (unfermented sap) into jaggery or sugar, the sap has gained increasing attention for its potential in various industries. Coconut inflorescence sap is packed with sugars, proteins, minerals, antioxidants, and vitamins, which the plant utilizes for growth and development of both tender and mature coconuts. The collection of unfermented sap has been a challenge, and the development of coco-sap chiller has revolutionized the way unfermented sap could be collected as nutritional beverage and processed into multiple, valuable food products (Hebbar et al., 2022; Sarma et al., 2023). Multiple entrepreneurs and coconut producer companies have adopted coco-sap chiller technology for collecting fresh sap and selling it as unfermented beverage (*Kalparasa*®) or processed into beverage packed by adopting TetraPak® technology or as coconut sugar. Because of its rich nutritive value, it is sold in the market as prime product. Ultra-high temperature treated *neera* is packed and marketed as '*Thenneera*' and has been exported from India to countries like USA, Europe and Middle East (https://thenneera.com/).

Coconut sap is classified as a low glycemic index (GI 35) food (Trinidad et al., 2015). However, another study reported a GI of 52.17, placing it in the intermediate range (Chinnamma et al., 2019). Variations in GI of coconut sap are known to be influenced by factors such as the age of palms, its genotype, climate, and geographical location. Thus, a comprehensive study is needed to accurately determine the GI of palm saps. However, the most appealing aspects of coconut sap are its flavour and nutritional content. This organic sap is known for its numerous health benefits, including anti-diabetic, anti-aging, anti-rheumatic, and anti-electrolyte imbalance properties (Hebbar et al., 2015).

Despite its high nutritional value and health benefits, its collection as a fresh beverage and marketing are highly unorganized. Otherwise, it was mainly consumed as a traditional drink in rural areas. With the advent of scientific cultivation methods, only in the last few years, it has been collected in organized manner. It is sold as fresh beverage in dispensers in local market, or minimally processed (pasteurization) and bottled and under refrigerated condition (with a shelf life of 15 days) or advanced packing (TetraPak®) (with a shelf life of 6 months). Technologies have been evolved to evaporate the sap into concentrate, jaggery or coconut sugar (Hebbar et al., 2022). A large-scale adoption of tapping and its sale as beverage or sap-based products revealed that tapping palms for inflorescence sap is 6 to 8-fold more profitable than selling the nuts. A palm which produces 100 nuts could yield as high as 400 l of sap. Glucose equivalent of coconut nuts and sap tapping is 38.5 and 133.0 kg, respectively suggesting tapping palms is three folds more energy efficient.

2. Challenges in the Sector

Thus, tapping of palms not only improves income of farmers but also create employment in terms of skilled tappers, processing personnel, marketing professionals etc., and also offer a quality product to the consumers. Inspite of these advantages, this sector still faces some challenges which are to be addressed to exploit its potential.

2.1. Climbing

A climber must reach the palm crown at least twice a day, in the morning and evening, to collect the gathered sap, slice the spadix, and reconnect the collection container. However, climbing tall palms poses significant risks, and the inherent danger involved reduces the likelihood of success in this type of work (Win et al., 2018). Unfortunately, there has been little automation in this process. The shortage of skilled climbers remains a major challenge in adopting efficient tapping techniques. Additionally, in many coconut growing countries, most orchards consist of senile and very tall trees, making climbing a cumbersome and risky task. Even when climbers are available, they charge exorbitant fees, significantly increasing the production cost of raw sap (*neera*). Consequently, the high production costs of processed products make them less competitive than juices or sugar derived from other sources in both domestic and international markets.

2.2. Tapping

Tapping or slicing the spadix for sap flow is a much more skilled task compared to harvesting tender or mature coconuts. While harvesting nuts can be done by climbing near the crown or the base of the crown, sap tapping is performed on the unopened inflorescence, which is typically located at the top of crown. Additionally, the spadix being tapped is bent and positioned almost perpendicular to the tree's axis, requiring the tapper to make precise cuts. Hence, this process demands advanced skills beyond those required for regular coconut harvesting.

2.3. Fermentation

Almost all coconut tree extracts, including sap, tender coconut water, mature coconut water, and coconut milk, are highly perishable. Due to its rich nutrient content, sap (*neera*) is particularly susceptible to enzymatic and microbial fermentation. Extensive efforts have been made to prevent sap deterioration through biological, chemical, physical, and thermal treatments. However, only thermal treatment has proven moderately effective. If the sap is collected and stored at temperatures around 2 to 4°C, it can retain its original form for 8 to 10 h. For long-term storage, sub-zero temperatures (- 20 to -40°C) are required. Therefore, maintaining a cold chain from collection to final processing is essential. This poses a significant challenge, particularly when coconut orchards are spread across distant locations, as it greatly increases costs.

2.4. Processing

Coconut sap in its original form has a neutral to alkaline pH, a characteristic shared by only a few natural beverages like water and milk. Unlike other juices, which are typically acidic and easier to preserve, coconut sap is highly prone to deterioration. When pasteurized at slightly higher temperatures, the sap tends to caramelize. Additionally, due to its sticky nature, it is challenging to use applications such as spray dryers or use various dryers to produce coconut sugar. Open-pan evaporation has proven to be the most successful method for drying the sap. Despite the challenges in collection, storage, preservation, and processing, coconut sap and its derived products offer notable health benefits.

3. Health and Wellness

3.1. Natural Sweetener

As consumers shift towards healthier lifestyles, coconut sap is seen as a low-GI natural sweetener alternative to sugar. Its use in health foods, diabetes-friendly products, and organic sweeteners could expand significantly. As stated already, coconut sap has a low GI, typically around 35, making it a preferred sweetener for diabetes (Trinidad et al., 2015). Coconut sap can be used as a direct substitute for sugar in various food products. It can sweeten baked goods, sauces, beverages, and desserts, offering a natural, healthier alternative to refined sugar and artificial sweeteners.

3.2. Natural Flavor Enhancer

Beyond sweetness, coconut sap has a unique flavor profile, with hints of caramel and butterscotch. This can enhance the taste of products in which it is used, providing an additional layer of complexity to foods and drinks.

3.3. Nutritional Supplements

Rich in vitamins, amino acids, and minerals, coconut sap could be marketed as a functional beverage or ingredient in nutritional supplements. Its potential for boosting immunity and providing natural energy makes it attractive in the wellness market. Unlike refined cane sugar, coconut sap is rich in vitamins (such as B vitamins), minerals (including potassium, magnesium, and zinc), and amino acids. These nutrients make it a more wholesome choice for consumers seeking not just sweetness but added nutritional benefits (Hebbar et al., 2020).

3.4. Food and Beverage Innovations

There is potential for the development of new products such as coconut sap-based syrups, candies, chocolates, and even frozen desserts. These products would appeal to health-conscious consumers who are looking for indulgence without compromising on health (Naik et al., 2015).

3.5. Fermented Products

The sap can be fermented to create a variety of products like coconut vinegar, coconut cider, and even low-alcohol beverages. These products align with the growing trend of fermented foods, which are valued for their probiotic content and digestive benefits (Flores-Gallegos et al., 2019).

3.6. Specialty Foods

There is potential to develop niche food products like coconut sap syrup, *neera*-based beverages, and even gourmet products because of its unique flavor profile and natural properties (Beegum et al., 2022).

3.7. Functional Foods

Coconut sap can be positioned as an ingredient in functional foods, that provide health benefits beyond basic nutrition. For example, it could be used in energy bars, smoothies, and other products targeted at athletes, and health enthusiasts.

3.8. Diet-Friendly Sweetener

Its low calorie and low carbohydrate content make coconut sap an ideal sweetener for various diets, including ketogenic, paleo, and low-carb diets. This versatility allows it to cater to a broad spectrum of dietary preferences (Ghosh et al., 2016).

4. Sustainability and Ethical Considerations

4.1. Eco-Friendly Practices

Harvesting coconut sap is considered sustainable and environmentally friendly, as it doesn't harm the tree and supports continuous production. Promoting this aspect could appeal to eco-conscious consumers. Coconut sap production is highly sustainable. The process of tapping coconut inflorescence does not harm the tree, allowing continuous production for many years. This contrasts with the environmental impact of other sweetener sources, such as sugar cane and corn.

4.2. Fair Trade Opportunities and Community Support

Ensuring fair wages and good working conditions for sap tappers could create fair trade-certified coconut sap products, adding value and market appeal. Ethical sourcing of coconut sap can support fair trade initiatives, ensuring that producers receive a fair wage. This not only benefits local communities but also enhances the product's appeal to socially conscious consumers.

5. Agricultural and Economic Benefits

5.1. Income Diversification for Farmers

For coconut farmers, tapping inflorescence sap provides an additional income stream alongside traditional coconut products like copra and coconut oil. This diversification can help stabilize incomes, especially in regions where coconut is a major crop.

5.2. Rural Development

Developing the *neera* industry can lead to job creation in rural areas, from sap collection to processing and marketing. It can also stimulate local economies by establishing cottage industries around sap-based products.

5.3. Clean Label Movement

Consumers are increasingly seeking out 'clean label' products—those with simple, natural, and easily recognizable ingredients. Coconut sap aligns perfectly with this trend, as it is minimally processed and free from synthetic additives and preservatives.

5.4. Organic and Non-GMO Markets

The demand for organic and non-GMO products is growing. Coconut sap can be marketed as an organic sweetener, especially if sourced from organic-certified farms. This opens up opportunities in premium market segments, and natural products.

6. The Way Forward

6.1. Climbing

Several institutions have been working to develop motorized climbing devices, robots, hydraulic masts, and other tools to assist with climbing. Currently, some climbers use ladders or manual paddle-type climbing machines, but these are not universally comfortable or practical. This underscores the need for further automation. Similarly, efforts are underway to develop microprocessor-controlled slicing and tapping machines to replace manual climbing. In the past, attempts were made to create a device that could substitute the manual massaging and stroking of the spadix, which is essential for sap flow. The integration of artificial intelligence (AI) and the Internet of Things (IoT) could potentially automate tasks such as climbing, tapping, and the collection of unfermented sap, significantly reducing labor costs and environmental impacts.

Recent advancements in the collection of coconut sap, its processing into by-products have led to a remarkable increase in both domestic and international markets. However, the coconut sap sugar value chain currently lacks traders or consolidators for coconut sap as a raw material, primarily due to low productivity rates, perishability of coconut sap, and challenges in coordinating sap harvesting and processing (Costales 2019). There is an enormous potential for improvement in various methods used for collecting coconut inflorescence sap beyond what has been developed so far.

To enhance this sector, there is a strong need for the development of technology to mechanize or automate tapping (Somawiharja et al., 2018), as well as for establishing dwarf and hybrid coconut palm plantations. Additionally, educational programs focusing on technology for tappers are essential. It is also crucial to create mechanized or automated systems that can collect sap without the use of preservatives. Research aimed at automating this process could be highly beneficial.

Furthermore, it is important to provide farmers and tappers with comprehensive information about coconut sap farming to optimize the use of resources effectively. Achieving these developments would represent a significant progress in coconut sap farming. Integrating advanced AI technologies (Concepcion et al., 2020; Concepcion et al., 2021; Palconit et al., 2021) would also offer substantial advantages.

A schematic diagram of the proposed potential solutions towards more efficient coconut sap production is given in Figure 1.

6.2. Varieties with Special Traits

Dwarf coconut trees are ideal for tapping due to their manageable height, though their sap yield is relatively low. However, an important constraint with dwarfs is the fragility of their leaves, which can easily break when a tapper climbs the palm, negatively impacting its growth. Therefore, breeding efforts are being made to develop dwarf varieties with canopies similar to tall trees but with stronger leaves. Hybrid palms such as dwarf x dwarf (D x D) or dwarf x tall (DxT) with medium height are particularly promising. Additionally, exploring the genetic and molecular characteristics of coconut trees could lead to innovations in sap yield, quality, and tapping methods, thereby improving production efficiency and product consistency. The use of fermented coconut sap as a starter culture for producing virgin coconut oil further expands the range of novel sap-based products (Pandiselvam et al.2023).

6.3. Product Stability and Shelf Life

Fresh coconut sap is prone to fermentation, which can affect its taste and usability as a sweetener. Ongoing research shall focus on improving the stability and shelf life of fresh coconut sap. Advances in preservation techniques could make it easier to store and transport, expanding its market reach. Research and innovation in preservation techniques, such as pasteurization, vacuum sealing, or drying into granules, can help extend its shelf life and make it more accessible to global markets (Ramesh et al., 2021).

6.4. Novel Product Development

Ongoing research into the properties and applications of coconut sap could lead to new product innovations. This might include developing different forms of the sweetener, such as powdered or crystallized versions, or enhancing its stability and flavor profile.

6.5. Nutritional Studies

Further research into the health benefits of coconut sap could boost its reputation as a super food sweetener. Clinical studies demonstrating its effects on blood sugar levels, weight management, and overall health could support its use in therapeutic diets.

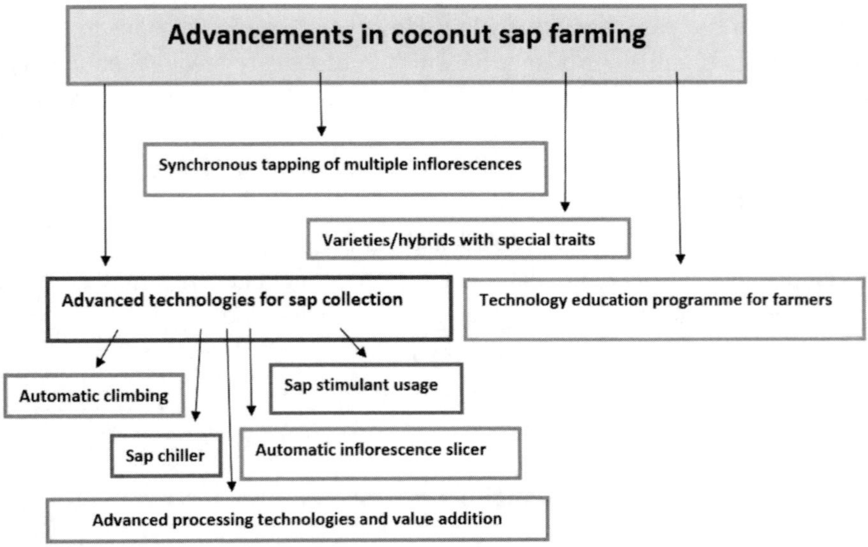

Figure 1. Potential solutions towards efficient coconut sap production.

6.6. Standardization

For coconut sap to succeed on a global scale, there needs to be standardization in terms of quality, purity, and labeling. Establishing these quality standards can help build consumer trust and meet international food safety regulations.

6.7. Market Awareness

Increasing consumer awareness about the benefits and uses of coconut sap is crucial. Effective marketing strategies and educational campaigns can help position it as a versatile and valuable product in the global market.

6.8. Price Competitiveness

Compared to conventional sugars, coconut sap sugar may be more expensive to produce and process. As production scales up and more efficient processing methods are developed, the cost could decrease, making it more competitive with other sweeteners.

Concluding Remarks

In conclusion, coconut inflorescence sap holds significant potential across various industries, driven by health trends, sustainability concerns, and economic benefits. With appropriate investment in research, development, and marketing, it could become a major player in the global natural products market. Coconut inflorescence sap has substantial potential as a natural sweetener in a health-conscious and sustainability-focused global market. Its nutritional advantages, versatility, and sustainable production without much use of chemical and pesticides make it a promising alternative to conventional beverages and sweeteners.

References

Alahmad, T., Neményi, M., Nyéki, A. (2023). Applying IoT sensors and big data to improve precision crop production: a review. *Agronomy* 13(10), 2603, https://doi.org/10.3390/agronomy13102603.

Beegum, P.P.S., Nair, J.P., Manikantan, M.R., Pandiselvam, R., Shil, S., Neenu, S., Hebbar, K.B. (2022) Effect of coconut milk, tender coconut and coconut sugar on the physico-chemical and sensory attributes in ice cream. *J Food Sci Technol* 59:2605–2616 https://doi.org/10.1007/s13197-021-05279-y.

Concepcion, II R., Dadios, E., Carpio, J., Bandala, A., Sybingco, E. (2020) Phytopigments profiling of *Lactuca sativa* leaf chloroplast photosystems via vision-based planar chromatography. In: *2020 IEEE 12th International Conference on Humanoid, Nanotechnology, Information Technology, Communication and Control, Environment, and Management (HNICEM),* 2020.

Concepcion, II R., Lauguico, S., Alejandrino, J., Bandala, A., Sybingco, E., Vicerra, R.R., Dadios, E., Cuello, J. (2021) Adaptive fertigation system using hybrid vision-based lettuce phenotyping and fuzzy logic valve controller towards sustainable aquaponics. *J Advanced Computational Intelligence and Intelligent Informatics* 25(5): 610-617, https://doi.org/10.20965/jaciii.2021.p0610.

Costales, (2019) 3 NTP Value chain analysis and development of VC models. *Excoll Travera Enterprise Study,* 2019.

Flores-Gallegos, A.C., Vázquez-Vuelvas, O.F., López-López, L.L., Sainz-Galindo, A., Ascacio- Valdes, J.A., Aguilar, C.N., Rodriguez-Herrera, R. (2019) Tuba, a fermented and refreshing beverage from coconut palm sap. In*: Non-alcoholic beverages.* Elsevier, pp 163–184

Ghosh, D.K., Bandyopadhyay, A., Baidyanath, D., Kaushik, G., Hebbar, K.B., Chowdappa, P. (2016) Method of the preparation of fresh coconut inflorescence sap (*Kalparasa*) based sweets. *Pat Appl* No 201631013187.

Hebbar, K.B., Arivalagan, M., Manikantan, M.R., Mathew, A.C., Thamban, C., Thomas, G.V., Chowdappa, P. (2015) Coconut inflorescence sap and its value addition as sugar–collection techniques, yield, properties and market perspective. *Curr Sci* 109(8):1411–1417.doi:10.18520/v109/i8/1411-1417.

Hebbar, K.B., Arivalagan, M., Pavithra, K.C., Roy, T.K., Gopal, M., Shivashankara, K.S., Chowdappa, P. (2020) Nutritional profiling of coconut (*Cocos nucifera* L.) inflorescence sap collected using novel coco-sap chiller method and its value-added products. *J Food Meas Charact* 14:2703–2712.

Hebbar, K.B., Ramesh, S. V., Ghosh, D.K., Beegum, P.P.S., Pandiselvam, R., Manikantan, M.R., Mathew, A.C. (2022) Coconut sugar-a potential storehouse of nutritive metabolites, novel bio-products and prospects. *Sugar Tech* 24(3):841–856.

Naik, A.C., Madhusudhan, M., Raghavarao, K., Subba, D. (2015) Downstream processing for production of value-added products from coconut. *Curr Biochem Eng* 2(2):168–180.

Palconit, M.G., Concepcion II R., Alejandrino, J., Pareja, M., Almero, V.J., Bandala, A., Vicerra, R.R., Sybingco, E., Dadios, E., Naguib, R. (2021) Three-dimensional stereo vision tracking of multiple free-swimming fish for low frame rate video. *J Advanced*

Computational Intelligence and Intelligent Informatics 25(5):639-646,doi: 10.20965/jaciii.2021.p0639.

Pandiselvam, R., Akshay, A.T., Manikantan, M.R., Ramesh, S. V., Patil, H., Gopal, M. (2023). Synergetic effect of fermented coconut inflorescence sap for the production of virgin coconut oil. *J Food Biochem* 2023(1):7125810.

Ramesh, S.V., Beegum, P.P.S., Sugatha, P., Selvamani, V., Pandiselvam, R., Manikantan, M.R., Gopal, M., Hebbar, K.B. (2021). *A Physical, biochemical, mineral, microbial and sensory test report on ultra high temperature (UHT) treated neera.* pp10. submitted to the client: M/s. Sky Agri Products, Tiruppur, Tamil Nadu – 641664.

Sarma, C., Mummaleti, G., Sivanandham, V., Kalakandan, S., Rawson, A., Anandharaj, A. (2022). Anthology of palm sap: The global status, nutritional composition, health benefits & value added products. *Trends Food Sci Technol* 119:530–549.

Somawiharja, Y., Purnomo, H., Wonohadidjojo, D.M., Kartikawati, M., Suniati, R.T. (2018) Indigenous technology of tapping, collecting and processing of coconut (*Cocos nucifera*) sap and its quality in Blitar Regency, East Java, Indonesia. *Food Res* 2(4): 398– 403, doi; 10.26656/fr.2017.2(4).075.

Sudha, R., Niral, V., Hebbar, K.B., Samsudeen, K. (2019). Coconut inflorescence sap. *Current Sci* 116(11):1809-1817, doi:10.18520/cs/v116/i11/1809-1817.

Win, T.H., Htura, Htun C. (2018) Factors in the Decreasing Number of Sap Tappers in Shwe-Hlaing Village, Nyaung U Township, Mandalay Region. *Universities Research Journal* 11(8):1-27.

About the Editors

K. B. Hebbar

Dr. K. B. Hebbar, currently serves as the Director of the ICAR-Central Plantation Crops Research Institute (CPCRI), Kasaragod, Kerala. A seasoned plant physiologist, Dr. Hebbar holds a Master's degree in Plant Physiology from the University of Agricultural Sciences, Bengaluru, and a PhD from the Indian Agricultural Research Institute, New Delhi, with specialization in Electrophysiology, Biochemistry, and Molecular Biology.

His career spans significant contributions to climate change research in plantation crops, focusing on drought, heat, salinity, and elevated CO_2 impacts on coconut, arecanut, and cocoa. As former Head of the Division of Physiology, Biochemistry, and Post-Harvest Technology at CPCRI, he led key research predicting future climatic suitability zones for these crops in India and developed innovative technologies such as the unfermented coconut sap tapping system for coconut sugar production.

Dr. Hebbar is a recipient of the Borlaug and Global Research Alliance Fellowship (2010), has authored over 100 high-impact publications, and holds multiple patents including for the coco sap chiller and snowball tender nut. His leadership extends to numerous national committees and technical advisory roles across Kerala, Karnataka, and Tamil Nadu. He has mentored many postgraduate and doctoral scholars and implemented R&D projects supported by national and international bodies, including ICAR, MoEF&CC, and IPCC.

V. Krishnakumar

Dr. V. Krishnakumar is a former Principal Scientist (Agronomy) & Head, ICAR-Central Plantation Crops Research Institute, Regional Station, Kayamkulam, Kerala. He holds Master's and Doctoral degrees in Agronomy and as an Agronomist, he has specialized in Soil Health Management and Crop Nutrition, Climate Resilient Agriculture, Crop Pluralism and Cropping System Management, and Organic Farming in Plantation Crops. As a member of the team, he has bagged the prestigious "*Outstanding Interdisciplinary*

Team Research in Agriculture and Allied Sciences for the Biennium 2005-06" instituted by the Indian Council of Agricultural Research. He is a recipient of the RL Narasimha Swamy Award and Fellow of the Indian Society for Plantation Crops.

He has functioned as the Principal Investigator/Co-Investigator for various externally funded projects of the Dept. of Biotechnology, Dept. of Science and Technology, Govt. of India; Coconut Development Board, GoI; Kerala State Planning Board, National Agricultural Technology Project-NATP & National Agricultural Innovation Project-NAIP, Government of India; and International Fund for Agriculture Development-IFAD/COGENT.

As a technical consultant, he has visited Guyana for the 'Support of the development of Guyana's coconut industry' and Mauritius for 'Capacity building for sustainable production and processing of coconut in Mauritius.' He has served as Editor of the *Journal of Plantation Crops* (ISPC) during 2006-2011. He has edited fourteen books and published more than 100 research and technical articles apart from contributing chapters in various books, technical bulletins, popular articles etc. He serves on the editorial boards of reputed journals including *Current Science, Journal of Plantation Crops, Journal of Spices and Aromatic Crops, and Asian Journal of Agricultural Extension, Economics and Sociology.*

dr.krishnavkumar@gmail.com.

S. V. Ramesh

Dr. S. V. Ramesh is a Senior Scientist (Biotechnology) at the ICAR-Central Plantation Crops Research Institute (ICAR-CPCRI), Kasaragod, Kerala, India. He holds Master's and Doctoral degrees in Plant Biochemistry and specializes in plant biochemistry and molecular biology. His post-doctoral research, conducted at Washington State University under the prestigious DBT-CREST Fellowship (2013–2014), focused on plant and viral small non-coding RNAs. He also underwent advanced training in legume bioinformatics at the J. Craig Venter Institute (JCVI), Maryland, USA.

Currently, Dr. Ramesh's research focuses on nutritional biochemistry and bioactives with nutraceutical potential in plantation crops. He has played a pivotal role in developing quality standards for virgin coconut oil and coconut testa oil and in advancing techniques for detecting adulteration in coconut oil. His contributions have been widely recognized, earning him accolades such as the DBT-CREST Award, DBT-CTEP Travel Award, AABSc Young Plant Biotechnologist Award, and DST-ITS Travel Award.

Dr. Ramesh has published extensively in reputed national and international journals, covering topics from plantation biology to plant-virus interactions. He has also edited six books and presented his research at numerous global scientific platforms including the World Soybean Research Conference (South Africa), APS-CPS Joint Meeting (USA), and the International Symposium on Cocoa Research (France). He serves on the editorial boards of reputed journals including *PLOS One*, *Plant Physiology Reports*, and the *Journal of Plantation Crops*.

Index

D

E

F

G

H

I

M

N

P

Q

R

S

T

U

V